THE FOUNDATIONS OF ANALYSIS:
A STRAIGHTFORWARD INTRODUCTION

Book 1
Logic, Sets and Numbers

THE FOUNDATIONS OF ANALYSIS: A STRAIGHTFORWARD INTRODUCTION

BOOK 1
LOGIC, SETS AND NUMBERS

K. G. BINMORE

Professor of Mathematics
London School of Economics and Political Science

CAMBRIDGE UNIVERSITY PRESS

Cambridge
London   New York   New Rochelle
Melbourne   Sydney

CAMBRIDGE UNIVERSITY PRESS
Cambridge, New York, Melbourne, Madrid, Cape Town, Singapore, São Paulo, Delhi

Cambridge University Press
The Edinburgh Building, Cambridge CB2 8RU, UK

Published in the United States of America by Cambridge University Press, New York

www.cambridge.org
Information on this title: www.cambridge.org/9780521233224

First published 1980
Re-issued in this digitally printed version 2008

A catalogue record for this publication is available from the British Library

ISBN 978-0-521-23322-4 hardback
ISBN 978-0-521-29915-2 paperback

CONTENTS

†This material is more advanced than the main body of the text and is perhaps best omitted at a first reading.

INTRODUCTION

This book contains an informal but systematic account of the logical and algebraic foundations of mathematical analysis written at a fairly elementary level. The book is entirely self-contained but will be most useful to students who have already taken, or are in the process of taking, an introductory course in basic mathematical analysis. Such a course necessarily concentrates on the notion of convergence and the rudiments of the differential and integral calculus. Little time is therefore left for consideration of the foundations of the subject. But the foundational issues are too important to be neglected or to be left entirely in the hands of the algebraists (whose views on what is important do not always coincide with those of an analyst). In particular, a good grasp of the material treated in this book is essential as a basis for more advanced work in analysis. The fact remains, however, that a quart will not fit into a pint bottle and only so many topics can be covered in a given number of lectures. In my own lecture course I deal with this problem to some extent by encouraging students to read the more elementary material covered in this book for themselves, monitoring their progress through problem classes. This seems to work quite well and it is for this reason that substantial sections of the text have been written with a view to facilitating 'self-study', even though this leads to a certain amount of repetition and of discussion of topics which some readers will find very elementary. Readers are invited to skip rather briskly through these sections if at all possible.

This is the first of two books with the common title

Foundations of Analysis: A Straightforward Introduction.

The current book, subtitled

Logic, Sets and Numbers,

was conceived as an introduction to the second book, subtitled

Topological Ideas

and as a companion to the author's previous book

Mathematical Analysis: A Straightforward Approach.

Although *Logic, Sets and Numbers* may profitably be read independently of these other books, I hope that some teachers will wish to use the three books together as a basis for a sequence of lectures to be given in the first two years of a mathematics degree.

Certain sections of the book have been marked with a † and printed in smaller type. This indicates material which, although relevant and interesting, I regard as unsuitable for inclusion in a first year analysis course, usually because it is too advanced, or else because it is better taught as part of an algebra course. This fact is reflected in the style of exposition adopted in these sections, much more being left to the reader than in the body of the text. Occasionally, on such topics as Zermelo–Fraenkel set theory or transfinite arithmetic, only a brief indication of the general ideas is attempted. Those reading the book independently of a taught course would be wise to leave those sections marked with a † for a second reading.

A substantial number of exercises have been provided and these should be regarded as an integral part of the text. The exercises are not intended as intelligence tests. By and large they require little in the way of ingenuity. and, in any case, a large number of hints are given. The purpose of the exercises is to give the reader an opportunity to test his or her understanding of the text. Mathematical concepts are sometimes considerably more subtle than they seem at first sight and it is often not until one has failed to solve some straightforward exercises based on a particular concept that one begins to realise that this is the case.

Finally, I would like to thank Mimi Bell for typing the manuscript for me so carefully and patiently.

June 1980   K. G. BINMORE

1 PROOFS

1.1 What is a proof?

Everyone knows that theorems require proofs. What is not so widely understood is the nature of the difference between a mathematical proof and the kind of argument considered adequate in everyday life. This difference, however, is an important one. There would be no point, for example, in trying to construct a mathematical theory using the sort of arguments employed by politicians when seeking votes.

The idea of a formal mathematical proof is explained in §1.5. But it is instructive to look first at some plausible types of argument which we shall *not* accept as proofs.

1.2 *Example* We are asked to decide whether or not the expression

$$n^3 - 4n^2 + 5n - 1$$

is positive for $n = 1, 2, 3, \ldots$. One approach would be to construct a table of the expression for as many values of n as patience allows.

n	$n^3 - 4n^2 + 5n - 1$
1	1
2	1
3	5
4	19
5	49
6	101

From the table it seems as though $n^3 - 4n^2 + 5n - 1$ simply keeps on getting larger and larger. In particular, it seems reasonable to guess that $n^3 - 4n^2 + 5n - 1$ is always positive when $n = 1, 2, 3, \ldots$. But few people would maintain that the argument given here is a *proof* of this assertion.

1.3 *Example* In this example the situation is not quite so clear. We

1

are asked to decide whether or not the expression

$$x^2 - 3x + 2$$

is negative for all values of x satisfying $1 < x < 2$. Since $x^2 - 3x + 2 = (x-1)(x-2)$, it is easy to draw a graph of the parabola $y = x^2 - 3x + 2$.

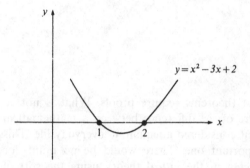

From the diagram it seems quite 'obvious' that $y = x^2 - 3x + 2$ is negative when $1 < x < 2$ and positive or zero otherwise. But let us examine this question more closely.

How do we know that the graph we have drawn really does represent the behaviour of the equation $y = x^2 - 3x + 2$? School children learn to draw graphs by plotting lots of points and then joining them up. But this amounts to guessing that the graph behaves as we think it should in the gaps between the plotted points. One might counter this criticism by observing that we know from our experience that the use of the graph always leads to correct answers. This would be a clinching argument in the field of physics. But, in mathematics, we are not supposed to accept arguments which are based on our experience of the world.

One might, of course, use a mathematical argument to deduce the properties of the graph, but then the graph would be unnecessary anyway.

We are forced (reluctantly) to the conclusion that an appeal to the graph of $y = x^2 - 3x + 2$ cannot be regarded as a *proof* that $x^2 - 3x + 2$ is negative for $1 < x < 2$.

1.4   *Example* In the diagram below, the point O has been chosen as the point of intersection of the bisector of the angle A and the perpendicular bisector of the side BC. The dotted lines are then constructed as shown.

With the help of this diagram we shall show by the methods of elementary geometry that $AB = AC$ – i.e. *all triangles are isosceles*.

The triangles AEO and AFO are congruent and hence

$$AE = AF \tag{1}$$

$$OE = OF. \tag{2}$$

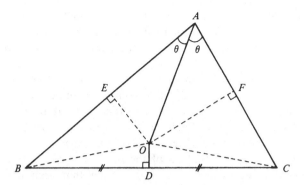

Also the triangles OBD and OCD are congruent. Thus

$$OB = OC. \tag{3}$$

From (2), (3) and Pythagoras' theorem it follows that

$$EB = FC. \tag{4}$$

Finally, from (1) and (4) we obtain that

$$AB = AC.$$

The explanation of this well-known fallacy is that the point O should lie *outside* the triangle ABC. In other words, the diagram does not represent the way things 'really are', and we have been led into error by depending on it. The objection that the diagram was not 'properly drawn' carries no weight since it is clearly not acceptable that a mathematical proof should depend on accurate measurement with ruler and compasses. This means, in particular, that the classical arguments of Euclidean geometry are not acceptable as proofs in modern mathematics because of their dependence on diagrams.

1.5   Mathematical proof

So far we have seen a number of arguments which are not proofs. What then is a proof?

Ideally, the description of a mathematical theory should begin with a list of *symbols*. This should be a finite list and contain all of the symbols which will be used in the theory. For even the simplest theory quite a few symbols will be needed. One will need symbols for variables – e.g. x, y and z. One will need symbols for the logical connectives (and, or, implies, etc.). The highly useful symbols) and (should not be forgotten and for a minimum of mathematical content one should perhaps include the symbols $+$ and $=$.

Having listed the symbols of the theory (and those mentioned above are just some of the symbols which might appear in the list), it is then necessary to specify how these symbols may be put together to make up *formulae* and then how such formulae may be put together to make up *sentences*.

Next, it is necessary to specify which of these sentences are to be called *axioms*.

Finally, we must specify *rules of deduction* which will tell us under what circumstances a sentence may be *deduced* from other sentences.

A mathematical *proof* of a *theorem* S is then defined to be a list of sentences, the last of which is S. Each sentence in the list must be either an *axiom* or else a *deduction* from sentences appearing earlier in the list.

What is more, we demand that all of the processes described above be specified so clearly and unambiguously that even that arch-idiot of intellectuals, the computer, could be programmed to check that a given list of sentences is a proof.

Of course, the ideas set out above only represent an ideal. It is one thing to set a computer to checking a list of several million sentences and quite another to prepare such a list for oneself. Apart from any other consideration, it would be extremely boring.

1.6   *Example* The list of sentences given below shows what a formal proof looks like. It is a proof taken from S. C. Kleene's *Introduction to Metamathematics* (North-Holland, 1967) of the proposition '$a = a$'. This does not happen to be one of his axioms and therefore needs to be proved as a theorem.

(1) $a = b \Rightarrow (a = c \Rightarrow b = c)$

(2) $0 = 0 \Rightarrow (0 = 0 \Rightarrow 0 = 0)$

(3) $\{a = b \Rightarrow (a = c \Rightarrow b = c)\} \Rightarrow \{[0 = 0 \Rightarrow (0 = 0 \Rightarrow 0 = 0)] \Rightarrow$
      $[a = b \Rightarrow (a = c \Rightarrow b = c)]\}$

(4) $[0 = 0 \Rightarrow (0 = 0 \Rightarrow 0 = 0)] \Rightarrow [a = b \Rightarrow (a = c \Rightarrow b = c)]$

(5) $[0 = 0 \Rightarrow (0 = 0 \Rightarrow 0 = 0)] \Rightarrow \forall c[a = b \Rightarrow (a = c \Rightarrow b = c)]$

(6) $[0 = 0 \Rightarrow (0 = 0 \Rightarrow 0 = 0)] \Rightarrow \forall b \forall c[a = b \Rightarrow (a = c \Rightarrow b = c)]$

(7) $[0 = 0 \Rightarrow (0 = 0 \Rightarrow 0 = 0)] \Rightarrow \forall a \forall b \forall c[a = b \Rightarrow (a = c \Rightarrow b = c)]$

(8) $\forall a \forall b \forall c[a = b \Rightarrow (a = c \Rightarrow b = c)]$

(9) $\forall a \forall b \forall c[a = b \Rightarrow (a = c \Rightarrow b = c)] \Rightarrow \forall b \forall c[a + 0 = b \Rightarrow (a + 0 = c \Rightarrow b = c)]$

(10) $\forall b \forall c[a + 0 = b \Rightarrow (a + 0 = c \Rightarrow b = c)]$

(11) $\forall b \forall c[a + 0 = b \Rightarrow (a + 0 = c \Rightarrow b = c)] \Rightarrow \forall c[a + 0 = a \Rightarrow (a + 0 = c \Rightarrow a = c)]$

(12) $\forall c[a + 0 = a \Rightarrow (a + 0 = c \Rightarrow a = c)]$

(13) $\forall c[a + 0 = a \Rightarrow (a + 0 = c \Rightarrow a = c)] \Rightarrow [a + 0 = a \Rightarrow (a + 0 = a \Rightarrow a = a)]$

(14) $a + 0 = a \Rightarrow (a + 0 = a \Rightarrow a = a)$

(15) $a + 0 = a$

(16) $a + 0 = a \Rightarrow a = a$

(17) $a = a$.

The above example is given only to illustrate that the formal proofs of even the most trivial propositions are likely to be long and tedious. What is more, although a computer may find formal proofs entirely satisfactory, the human mind needs to have some explanation of the 'idea' behind the proof before it can readily assimilate the details of a formal argument.

What mathematicians do in practice therefore is to write out 'informal proofs' which can 'in principle' be reduced to lists of sentences suitable for computer ingestion. This may not be entirely satisfactory, but neither is the dreadfully boring alternative. In this book our approach will be even less satisfactory from the point of view of those seeking 'absolute certainty', since we shall not even describe in detail the manner in which mathematical assertions can be coded as formal lists of symbols. We shall, however, make a serious effort to remain true to the spirit of a mathematical proof, if not to the letter.

1.7   Obvious

The word 'obvious' is much abused. We shall follow the famous English mathematician G. H. Hardy in interpreting the sentence '*P* is obvious' as meaning 'It is easy to think of a proof of *P*'. This usage accords with what was said in the section above.

A much more common usage is to interpret '*P* is obvious' as meaning 'I *cannot* think of a proof of *P* but I am sure it must be true'. This usage should be avoided.

1.8   The interpretation of a mathematical theory

Observe that in our account of a formal mathematical theory the content has been entirely divorced from 'reality'. This is so that we can be sure, in so far as it is possible to be sure of anything, that the theorems are correct.

But mathematical theories are not made up at random. Often they arise as an attempt to abstract the essential features of a·'real world' situation. One sets up a system of axioms each of which corresponds to a well-established 'real world' fact. The theorems which arise may then be interpreted as predictions about what happens in the 'real world'.

But this viewpoint can be reversed. In many cases it turns out to be very useful when seeking a proof of a theorem to think about the real world situation of which the mathematical theory is an abstraction. This can often suggest an approach which might not otherwise come to mind. It is sometimes useful, for example, to examine theorems in complex analysis in terms of their electrostatic interpretation. In optimisation theory, insight can sometimes be obtained by viewing the theorems in terms of their game-theoretic or economic interpretation.

For our purposes, however, it is the interpretation in terms of geometry that we shall find most useful. One interprets the real numbers as points along an ideal ruler with which we measure distances in Euclidean geometry. This interpretation allows us to draw pictures illustrating propositions in analysis. These pictures then often suggest how the theorem in question may be proved. But it must be emphasised again that these pictures cannot serve as a *substitute* for a proof, since our theorems should be true regardless of whether our geometric interpretation is a good one or a bad one.

2 LOGIC (I)

2.1 Statements

The purpose of logic is to label sentences either with the symbol T (for *true*) or with the symbol F (for *false*). A sentence which can be labelled in one of these two ways will·be called a *statement*.

2.2 *Example* The following are both statements.

(i) Trafalgar Square is in London.
(ii) $2+2=5$.

The first is true and the second is false.

2.3 *Exercise*

Which of the following sentences are statements?

(i) More than 10 000 000 people live in New York City.
(ii) Is Paris bigger than Rome?
(iii) Go jump in a lake!
(iv) The moon is made of green cheese.

2.4 Equivalence

From the point of view of logic, the only thing which really matters about a statement is its *truth value* (i.e. T or F). Thus two statements P and Q are *logically equivalent* and we write

$$P \Leftrightarrow Q$$

if they have the same truth value. If P and Q are both statements, then so is $P \Leftrightarrow Q$ and its truth value may be determined with the aid of the following *truth table*.

P	Q	$P \Leftrightarrow Q$
T	T	T
T	F	F
F	T	F
F	F	T

In this table the right-hand column contains the truth value of '$P \Leftrightarrow Q$' for all possible combinations of the truth values of the statements P and Q.

2.5   *Example* Let P denote the statement 'Katmandu is larger than Timbuktu' and Q denote the statement 'Timbuktu is smaller than Katmandu'. Then P and Q are logically equivalent even though it would be quite difficult in practice to determine what the truth values of P and Q are.

2.6   Not

If P is a statement, the truth value of the statement (not P) may be determined from the following truth table.

P	not P
T	F
F	T

2.7   And, or

If P and Q are statements, the statements 'P and Q' and 'P or Q' are defined by the following truth tables.

P	Q	P and Q	P	Q	P or Q
T	T	T	T	T	T
T	F	F	T	F	T
F	T	F	F	T	T
F	F	F	F	F	F

The English language is somewhat ambiguous in its use of the word 'or'. Sometimes it is used in the sense of 'either/or' and sometimes in the sense of 'and/or'. In mathematics it is always used in the second of these two senses.

2.8   *Example* Let P be the statement 'The Louvre is in Paris' and Q the statement 'The Kremlin is in New York City'. Then 'P and Q' is false but '*P* and (not *Q*)' is true. On the other hand, '*P* or *Q*' and '*P* or (not *Q*)' are both true.

2.9   *Exercise*

(1) If *P* is a statement, (not *P*) is its *contradictory*. Show by means of truth tables that '*P* or (not *P*)' is a *tautology* (i.e. that it is true regardless of the truth or falsehood of *P*). Similarly, show that '*P* and (not *P*)' is a *contradiction* (i.e. that it is false regardless of the truth or falsehood of *P*).

(2) The following pairs of statements are equivalent regardless of the truth or falsehood of the statements *P*, *Q* and *R*. Show this by means of truth tables in the case of the odd numbered pairs.

   (i) *P*, not (not *P*)
   (ii) *P* or (*Q* or *R*), (*P* or *Q*) or *R*
   (iii) *P* and (*Q* and *R*), (*P* and *Q*) and *R*
   (iv) *P* and (*Q* or *R*), (*P* and *Q*) or (*P* and *R*)
   (v) *P* or (*Q* and *R*), (*P* or *Q*) and (*P* or *R*)
   (vi) not (*P* and *Q*), (not *P*) or (not *Q*)
   (vii) not (*P* or *Q*), (not *P*) and (not *Q*).

[*Hint*: For example, the column headings for the truth table in (iii) should read

P	*Q*	*R*	*P* and *Q*	(*P* and *Q*) and *R*	*Q* and *R*	*P* and (*Q* and *R*)

There should be *eight* rows in the table to account for all the possible truth value combinations of *P*, *Q* and *R*.]

(3) From 2(ii) above it follows that it does not matter how brackets are inserted in the expressions '*P* or (*Q* or *R*)' and '(*P* or *Q*) or *R*' and so we might just as well write '*P* or *Q* or *R*'. Equally we may write '*P* and *Q* and *R*' instead of the statements of 2(iii).

   Show by truth tables that the statements '(*P* and *Q*) or *R*' and '*P* and (*Q* or *R*)' need not be equivalent.

(4) Deduce from 2(iv) that the statements '*P* and (*Q*₁ or *Q*₂ or *Q*₃)', '(*P* and *Q*₁) or (*P* and *Q*₂) or (*P* and *Q*₃)' are equivalent. Write down similar results which arise from 2(v), 2(vi) and 2(vii). What happens with four or more *Q*s?

2.10   Implies

   Suppose that *P* and *Q* are two statements. Then the statement '*P* implies *Q*' (or '*P* ⇒ *Q*') is defined by the following truth table.

P	Q	P implies Q
T	T	T
T	F	F
F	T	T
F	F	T

In simple terms the truth of '*P* implies *Q*' means that, from the truth of *P*, we can deduce the truth of *Q*. In English this is usually expressed by saying

'*If P, then Q*'

or sometimes

'*P is a sufficient condition for Q*'.

It strikes some people as odd that '*P* implies *Q*' should be defined as true in the case when *P* is false. This is so that a proposition like '$x > 2$ implies $x > 1$' may be asserted to be true for all values of x.

Consider next the following truth table.

P	Q	P implies Q	not Q	not P	(not Q) implies (not P)
T	T	T	F	F	T
T	F	F	T	F	F
F	T	T	F	T	T
F	F	T	T	T	T

Observe that the entries in the third and sixth columns are identical. This means that the statements '*P* implies *Q*' and '(not *Q*) implies (not *P*)' are either *both* true or *both* false – i.e. they are logically equivalent. Wherever we see '*P* implies *Q*' we might therefore just as well write '(not *Q*) implies (not *P*)' since this is an equivalent statement.

The statement '(not *Q*) implies (not *P*)' is called the *contrapositive* of '*P* implies *Q*'. In ordinary English it is usually rendered in the form

'*P only if Q*'

or sometimes

'*Q is a necessary condition for P*'.

These last two expressions are therefore two more paraphrases for the simple statement '*P* implies *Q*'.

2.11   *Exercise*

(1) Given that the statements '*P*' and '$P \Rightarrow Q$' are both true, deduce that *Q* is true. [*Hint*: Delete from the truth table for $P \Rightarrow Q$ those rows which do

not apply.] If, instead, you are given that the statements '*Q*' and '*P⇒Q*' are true, show that *no* conclusion may be drawn about the truth value of *P* (unless further information is available).

[*Note*: The first rule of deduction described above is called 'Modus Ponens'. To deduce *P* from '*Q*' and '*P⇒Q*' has been called 'Modus Morons'.]

(2) Given that *P⇒Q* is true but *Q* false, show that *P* is false.

(3) The great philosopher Descartes based his system of philosophy on the principle 'cogito ergo sum' (I think, therefore I am). Dr Strabismus (whom God preserve) of Utrecht has founded a rival system based on the principle 'Non cogito ergo non sum' (I do not think, therefore I am not).

   Express both these principles in the form '*P⇒Q*' and hence clear Dr Strabismus (and the *Daily Express*, where his thoughts were once published) from the insinuation that his principle may be deduced from that of Descartes.

(4) Rain on Tuesday is a necessary condition for rain on Sunday. If it rains on Tuesday then it rains on Wednesday. But it rains on Wednesday only if it rains on Friday. Moreover no rain Monday implies no rain on Friday. Finally, rain on Monday is a sufficient condition for rain on Saturday.

   Express each of these statements in the form '*P⇒Q*'. Given that it rains on Sunday, what can be said about Saturday's weather?

(5) Show that the statement '*P* implies *Q*' is equivalent to '(not *P*) or *Q*'. Deduce that '*P* implies *Q*' is also equivalent to 'not {*P* and (not *Q*)}'.

(6) What conclusion (if any) may be drawn from the truth of '(not *P*)⇒*P*'?

2.12 If and only if
Consider the truth table below.

P	*Q*	*P⇒Q*	*Q⇒P*	(*P⇒Q*) and (*Q⇒P*)	*P⇔Q*
T	*T*	*T*	*T*	*T*	*T*
T	*F*	*F*	*T*	*F*	*F*
F	*T*	*T*	*F*	*F*	*F*
F	*F*	*T*	*T*	*T*	*T*

Observe that the entries in the fifth and sixth columns are identical. Thus to say that '*P* is logically equivalent to *Q*' is the same thing as saying '(*P* implies *Q*) and (*Q* implies *P*)'.

In ordinary English this statement is usually expressed in the form

'*P if and only if Q*'

– i.e. '(P only if Q) and (if Q then P)'. Alternatively it may be paraphrased as 'P is a necessary and sufficient condition for Q'.

2.13   Proof schema

Most theorems take one of the two forms 'If P, then Q' or 'P if and only if Q'. The second of these really consists of two theorems in one and is usually proved in two parts. First it is shown that 'P implies Q' and then that 'Q implies P'.

To show that 'P implies Q' we normally use one of the following three methods.

(1) The most straightforward method is to assume that P is true and try and deduce the truth of Q (there is no need to worry about what happens when P is false since then 'P implies Q' is automatically true).

(2) A second method is to write down the contrapositive '(not Q) implies (not P)' and prove this instead. We then assume the truth of (not Q) and try and deduce the truth of (not P).

(3) Finally, we may argue by contradiction (*reductio ad absurdum*). For this argument we assume the truth of *both* P *and* (not Q) and seek a contradiction.

From this we conclude that our hypothesis 'P and (not Q)' is false – i.e. 'not (P and (not Q))' is true. But this is equivalent to 'P implies Q' (exercise 2.11(5)).

In practice Q may often be quite a complicated statement. In this case the effectiveness of the second and third methods depends on the extent to which the contradictory of Q (i.e. not Q) can be expressed in a useful form. In general, it is a good idea to get the 'not' as far inside the expression as possible. For example, 'not $\{(P$ and $Q)$ or (not $R)\}$' is more usefully expressed in the form '$\{($not $P)$ or (not $Q)\}$ and R'. We return to this topic again in the next chapter.

2.14   *Example* We give three different proofs of the fact that, if $x^2 - 3x + 2 < 0$, then $x > 0$.

Proof 1 Assume that $x^2 - 3x + 2 < 0$. Then

$$3x > x^2 + 2 \geqq 2 \quad \text{(because } x^2 \geqq 0\text{)}.$$

Hence $x > \frac{2}{3} > 0.$

It follows that '$x^2 - 3x + 2 < 0$ implies $x > 0$'.

Proof 2 The contrapositive statement is that '$x \leqq 0$ implies $xx^2 - 3x + 2 \geqq 0$'. We therefore assume that $x \leqq 0$. Then

$$x - 1 \leqq 0 \quad \text{and} \quad x - 2 \leqq 0.$$

Hence $$x^2 - 3x + 2 = (x-1)(x-2) \geqq 0.$$

It follows that '$x \leqq 0$ implies $x^2 - 3x + 2 \geqq 0$' and hence that '$x^2 - 3x + 2 < 0$ implies $x > 0$'.

Proof 3 Assume $x^2 - 3x + 2 < 0$ *and* $x \leqq 0$. Then

$$x^2 < 3x - 2 \leqq -2 < 0.$$

This is a contradiction and hence '$x^2 - 3x + 2 < 0$ implies $x > 0$'.

3 LOGIC (II)

3.1 Predicates and sets

A *set* is a collection of objects which are called its *elements*. If a is an element of the set S, we say that a *belongs to* S and write

$$a \in S.$$

If b does not belong to S, we write $b \notin S$.

In a given context, there will be a set to which all the objects we wish to consider belong – i.e. the set over which we want our variables to range. This set is called the *universal set U*. It is important to emphasise that the universal set will not always be the same. If we are discussing the properties of the real number system, we shall want U to be the set of all real numbers. If we are discussing people, we shall want U to be the set of all human beings.

When it is clearly understood what the universal set is, we can discuss *predicates*. A predicate is a sentence which contains one or more variables but which becomes a statement when we replace the variables by objects from their range.

3.2 *Examples*

(i) Let the universal set be the set $\mathbb{R}$ of all real numbers. Some examples of predicates which are meaningful for this universal set are:

(a) $x > 3$ (b) $2x = x^2$ (c) $x + y = z$.

These predicates can be converted into statements by replacing x, y and z by objects from their range. For example, the substitutions $x = 2$, $y = 1$, $z = 3$ yield the statements

(a) $2 > 3$ (b) $2 \cdot 2 = 2^2$ (c) $2 + 1 = 3$

of which the first is false and the others true.

(ii) Let the universal set be the set of all people. Some examples of predicates which are meaningful for this universal set are:

(a) x is president   (b) x loves y.

If x and y are replaced by Romeo and Juliet respectively, we obtain two statements of which the first is false and the second true.

One use of predicates is in specifying sets. The simplest way of specifying a set is by listing its elements. We use the notation

$$A = \{\tfrac{1}{2}, 1, \sqrt{2}, e, \pi\}$$

to denote the set whose elements are the real numbers $\tfrac{1}{2}$, 1, $\sqrt{2}$, e and π. Similarly,

$$B = \{\text{Romeo, Juliet}\}$$

denotes the set whose elements are Romeo and Juliet. But this notation is, of course, no use in specifying a set which has an infinite number of elements. For such sets one has to name the property which distinguishes elements of the set from objects which are not in the set. Predicates are used for this purpose. For example, the notation

$$C = \{x : x > 3\}$$

denotes the set of all real numbers larger than 3 provided that it is understood that the variable x ranges over the universal set $\mathbb{R}$. Similarly,

$$D = \{y : y \text{ loves Romeo}\}$$

denotes the set of people who love Romeo, provided that the variable y is understood to range over the universal set of all people.

It is convenient to have a notation for the *empty* set $\emptyset$. This is the set which has *no* elements. For example, if x ranges over the real numbers, then

$$\{x : x^2 + 1 = 0\} = \emptyset.$$

This is because there are no real numbers x such that $x^2 = -1$.

3.3   *Exercise*

(1) Let the universal set U be the set $\mathbb{N}$ of all natural numbers (i.e. 1, 2, 3, ...). Then, for example, $\{x : 2 \leq x \leq 4\} = \{2, 3, 4\}$. Obtain similar identities in each of the following cases.

   (i) $\{x : x^2 - 3x + 2 = 0\}$ (ii) $\{x : x^2 - x + 2 = 0\}$
   (iii) $\{x : x^2 + 3x + 2 = 0\}$ (iv) $\{x : x^2 \leq 3 + x\}$.

(2) Let the universal set U be the set $\mathbb{R}$ of all real numbers. If

$$S = \{x : x^2 - 3x + 2 \leq 0\},$$

decide which of the following statements are true:

(i) $1 \in S$ (ii) $2 \notin S$ (iii) $2 \in S$ (iv) $4 \notin S$.

3.4   Quantifiers

A predicate may be converted into a statement by substituting for the variables objects from their range. For example, the predicate '$x > 3$' becomes the false statement '$2 > 3$' when 2 is substituted for x.

Another way that predicates may be converted into statements is with the use of the quantifiers *'for any'* and *'there exists'*.

For example, from the predicates '$x > 3$' and 'x loves y' we may obtain the statements

'For any x, $x > 3$'

'For any y, there exists an x such that x loves y'.

These statements may be paraphrased in the more natural forms 'All real numbers are bigger than three' and 'Everybody can find somebody to love them'.

It is sometimes convenient to use the abbreviations '$\forall$' instead of 'for any' and '$\exists$' instead of 'there exists'. With this notation the statements above become

$$\forall x(x > 3)$$
$$\forall y \, \exists x(x \text{ loves } y).$$

3.5   *Exercise*

(1) Paraphrase in ordinary English the following statements. Venture an opinion in each case as to whether the given statement is true or false.

(i) $\exists x(x > 3)$ (ii) $\forall x(2x = x^2)$ (iii) $\forall x \forall z \exists y(x + y = z)$
(iv) $\exists x(x \text{ is President})$ (v) $\exists x \forall y(x \text{ loves } y)$.

(2) Examine carefully the meaning of the two statements

(i) $\forall y \exists x(x < y)$ (ii) $\exists x \forall y(x < y)$

and explain why they are *not* the same.

(3) Introduce the abbreviations $\forall$ and $\exists$ into the following sentence:

'For any number x, we can find a number z such
that for every number y, $xy = z$'.

(4) If the universal set U is the set of all natural numbers, find the elements

of the sets

(i) $\{x : \forall y(xy = y)\}$ (ii) $\{x : \exists y(xy = 12)\}$.

3.6   Manipulations with quantifiers

The contradictory of the statement '$\forall x(x > 3)$' is simply 'not $\forall x(x > 3)$'. In the previous chapter we noted that it is desirable to get the 'not' as far inside the expression as possible. How does one go about this when quantifiers are involved?

In the example given above, the variable x ranges over an infinite set, namely the set of all real numbers. It is instructive to consider first a case where the variable ranges over only a finite set.

3.7   *Example* Let the universal set be the set of all human beings and consider the statement

$$\forall x(\text{God loves } x).$$

This means the same as

'(God loves Romeo) *and* (God loves Juliet) *and* ...'

where dots indicate that we continue until we have listed all human beings. From chapter 2 (exercise 2.9(4)) we know how to write down the contradictory of this statement. It is

'(God does not love Romeo) *or* (God does not love Juliet)

or ...'.

That is to say, 'There exists at least one person whom God does not love' – i.e.

$$\exists x\{\text{not (God loves } x)\}.$$

This mode of reasoning leads us to the rules embodied in the table below.

Statement	Contradictory
$\forall x\ P(x)$	$\exists x(\text{not } P(x))$
$\exists x\ P(x)$	$\forall x(\text{not } P(x))$

As we have seen above, these rules only have something new to say in the case when x ranges over an *infinite* set.

3.8   *Example*

Statement	Contradictory
$\forall x(x>3)$	$\exists x(x\leq 3)$
$\exists x(2x=x^2)$	$\forall x(2x\neq x^2)$
$\exists x\ \forall y(x$ loves $y)$	$\forall x\ \exists y(x$ does not love $y)$
$\forall x\ \forall z\ \exists y(x+y=z)$	$\exists x\ \exists z\ \forall y(x+y\neq z)$

The third of these may be obtained in two stages. First one notes that 'not $\exists x\ \forall y(x$ loves $y)$' is equivalent to '$\forall x$ {not $\forall y(x$ loves $y)$}' which is in turn equivalent to '$\forall x\ \exists y$ {not $(x$ loves $y)$}'.

3.9   *Exercise*

(1) Write down the contradictories of the following statements in a useful form.

(i) $\forall y\ \exists x(x<y)$ (ii) $\exists x\ \forall y(x<y)$
(iii) $\exists x(x>3$ and $x=4)$
(iv) Everybody can find someone to love them.

(2) Apart from the rules for forming contradictories discussed above there are some other rules for use with quantifiers. The table below contains a list of equivalent statements.

Q and $(\forall x\ P(x))$	$\forall x(Q$ and $P(x))$
Q or $(\forall x\ P(x))$	$\forall x(Q$ or $P(x))$
Q and $(\exists x\ P(x))$	$\exists x(Q$ and $P(x))$
Q or $(\exists x\ P(x))$	$\exists x(Q$ or $P(x))$

Justify the third of these equivalences in the case when x ranges over a finite set.

3.10   **More on contradictories**

We often have to form the contradictory of a statement like

'For any $x>0$, $x^2+x-2>0$'.

This tells us that, for the purposes of this statement, the range of the variable x is the set of *positive* real numbers instead of the set of all real numbers. This makes no difference to the way in which we form the contradictory, except that we must remember to keep the range of the variable the *same* in the contradictory as it is in the statement.

Thus the contradictory of the statement above is

'There exists an $x>0$ such that $x^2+x-2\leq 0$'.

3.11   *Examples*

(i) The contradictory of the statement 'There exists an $x<1$ such that $x^2=1$' is 'For any $x<1$, $x^2\neq 1$'.

(ii) The contradictory of the statement 'For any $x<4$ there exists a $y>2$ such that $x+y=3$' is 'There exists an $x<4$ such that for any $y>2$, $x+y\neq 3$'.

3.12   *Exercise*

(1) Write down in a useful form the contradictories of the following statements:

   (i) There exists an $x<3$ such that $x>2$.
   (ii) For any x satisfying $0<x<1$, $x^2<x$.
   (iii) For any $\varepsilon>0$ there exists a $\delta>0$ such that for any x satisfying $0<x<\delta$, $x^2<\varepsilon$ (i.e. $x^2\to 0$ as $x\to 0+$).
   (iv) Some Montague is hated by every Capulet.

3.13   Examples and counter-examples

To prove a statement of the type '$\forall x(P(x))$' can be quite demanding. One has to give an argument which demonstrates the truth of $P(x)$ *whatever* the value of x. It is certainly not enough to give some examples of values of x for which it can be seen that $P(x)$ is true.

To disprove a statement of the type '$\forall x(P(x))$' is much easier. This is the same as proving '$\exists x(\text{not } P(x))$' and to do this one need only find a single y for which $P(y)$ is false. We say that such a y provides a *counter-example* to the statement $\forall x(P(x))$.

3.14   *Example* Show that the statement

'For any $x>0$, $x^2-3x+2\geq 0$'

is false. A single counter-example will suffice and an appropriate value of x is $\frac{3}{2}$. We have that

$$(\tfrac{3}{2})^2-3(\tfrac{3}{2})+2=-\tfrac{1}{4}<0.$$

3.15   *Example* Show that it is false that α^β is irrational for all positive irrational real numbers α and β.

Either $\alpha_1 = \beta_1 = \sqrt{2}$ provides a counter-example, or else

$$\alpha_2 = (\sqrt{2})^{\sqrt{2}}$$

is irrational. In the latter case, take $\beta_2 = \sqrt{2}$. Then

$$\alpha_2{}^{\beta_2} = \{(\sqrt{2})^{\sqrt{2}}\}^{\sqrt{2}} = (\sqrt{2})^2 = 2$$

and hence α_2 and β_2 provide a counter-example.

This example is somewhat unusual in being non-constructive, i.e. we demonstrate that a counter-example exists without being able to say which of (α_1, β_1) and (α_2, β_2) it is.

4 SET OPERATIONS

4.1 Subsets

If S and T are sets, we say that S is a *subset* of T and write

$$S \subset T$$

if every element of S is also an element of T – i.e. $x \in S$ implies $x \in T$.

4.2   *Example* Let $A = \{1, 2, 3, 4\}$ and let $B = \{2, 4\}$. Then $B \subset A$. Notice that this is *not* the same thing as writing $B \in A$ (i.e. B is an element of A). The elements of A are simply 1, 2, 3 and 4 and B is not one of these.

It is often convenient to illustrate the relations which hold between sets by means of a diagram (a *Venn diagram*). The diagram drawn illustrates the

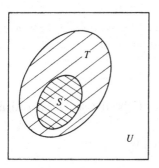

relation $S \subset T$. The universal set is represented by the square and the sets S and T by the shaded regions within the square.

4.3   *Exercise*

(1) Arrange the sets given in exercise 3.3(1) in a list, each item of which is a subset of the item which follows it.

(2) Draw a Venn diagram to illustrate the relation 'S is not a subset of T'.

(3) Justify the following

(i) $S \subset U$ (ii) $\emptyset \subset S$ (iii) $S \subset S$.

[*Hint*: Recall that $\emptyset$ denotes the empty set. Thus, for any x, $x \in \emptyset$ is *false* and therefore implies anything!]

(4) Prove that $S = T$ if and only if $S \subset T$ and $T \subset S$. [*Hint*: $S = T$ means '$x \in S \Leftrightarrow x \in T$'.]

4.4   Complements

If S is a set, its complement $\mathcal{C}S$ is defined by

$$S = \{x : x \notin S\}.$$

In the accompanying diagram, $\mathcal{C}S$ is represented by the shaded region.

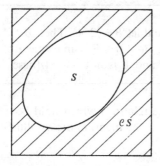

When using the above notation for the complement of a set, it is obviously important to be very clear about what universal set U is being used.

4.5   *Example* Let the universal set U be the set of all real numbers and let $S = \{x : x > 0\}$. Then

$$\mathcal{C} S = \{x : x \leq 0\}.$$

4.6   *Exercise*

(1) Let $U = \{1, 2, 3, 4, 5, 6\}$. Write down the complements of the sets $\{1, 2\}$ and $\{1, 3, 5\}$.

(2) Prove the following.

(i) $\mathcal{C}(\mathcal{C}S) = S$ (ii) $\mathcal{C}U = \emptyset$ (iii) $\mathcal{C}\emptyset = U$.

4.7 Unions and intersections

Suppose that S and T are sets. Their *union* $S \cup T$ and their *intersection* $S \cap T$ are defined by

$$S \cup T = \{x : x \in S \text{ or } x \in T\}$$
$$S \cap T = \{x : x \in S \text{ and } x \in T\}.$$

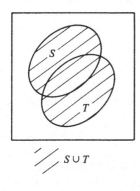 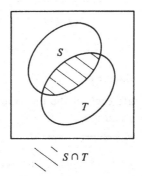

4.8   *Examples*

(i) $\{1, 2\} \cup \{1, 3, 5\} = \{1, 2, 3, 5\}$
(ii) $\{1, 2\} \cap \{1, 3, 5\} = \{1\}$.

4.9   *Exercise*

(1) We use the notation $A \setminus B$ to denote the set of all elements of A which do not belong to B – i.e. $A \setminus B = A \cap \mathcal{C} B$. Indicate the set $A \setminus B$ on a Venn diagram.

(2) Two sets A and B are called *disjoint* if they have no elements in common – i.e. $A \cap B = \emptyset$. Illustrate this situation on a Venn diagram.

(3) Use exercise 2.9(2) to obtain the following identities.

(i) $A \cap (B \cap C) = (A \cap B) \cap C$
(ii) $A \cup (B \cup C) = (A \cup B) \cup C$
(iii) $A \cup (B \cap C) = (A \cup B) \cap (A \cup C)$
(iv) $A \cap (B \cup C) = (A \cap B) \cup (A \cap C)$
(v) $\mathcal{C}(A \cap B) = \mathcal{C} A \cup \mathcal{C} B$
(vi) $\mathcal{C}(A \cup B) = \mathcal{C} A \cap \mathcal{C} B$.

Results like those of exercise 4.9(3) are conveniently illustrated by means of Venn diagrams. For example, problem (vi) may be illustrated as below.

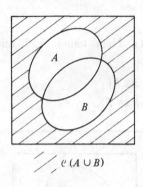
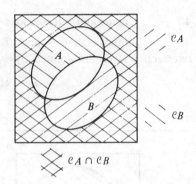

$$\mathcal{C}(A \cup B)$$ $$\mathcal{C}A \cap \mathcal{C}B$$

4.10   *Exercise*

(1) Illustrate each of the identities of exercise 4.9(3) by means of a pair of Venn diagrams.

So far we have considered unions and intersections of pairs of sets. We can equally well take the union or intersection of a whole collection of sets.

Suppose that $\mathcal{W}$ denotes a collection of subsets of a universal set U. Then we define

$$\bigcup_{S \in \mathcal{W}} S = \{x : \text{there exists an } S \in \mathcal{W} \text{ such that } x \in S]$$

$$\bigcap_{S \in \mathcal{W}} S = \{x : \text{for any } S \in \mathcal{W}, \ x \in S\}.$$

4.11   *Examples*

(i) Suppose that $\mathcal{W}$ contains only three sets, P, Q and R – i.e. $\mathcal{W} = \{P, Q, R\}$. Then

$$\bigcup_{S \in \mathcal{W}} S = P \cup Q \cup R \qquad \bigcap_{S \in \mathcal{W}} S = P \cap Q \cap R.$$

The appropriate Venn diagrams are

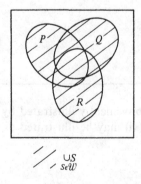
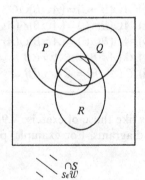

$$\bigcup_{S \in \mathcal{W}} S$$ $$\bigcap_{S \in \mathcal{W}} S$$

(ii) Let $\mathcal{W}$ be any collection of subsets of a universal set U. Prove that

$$\mathcal{C}\left(\bigcup_{S\in\mathcal{W}} S\right)=\bigcap_{S\in\mathcal{W}} \mathcal{C} S.$$

(See exercise 4.9(3vi).)

Proof $\mathcal{C}\left(\bigcup_{S\in\mathcal{W}} S\right)=\left\{x:\text{not}\left(x\in\bigcup_{S\in\mathcal{W}} S\right)\right\}$

$=\{x:\text{not (there exists an } S\in\mathcal{W} \text{ such that } x\in S)\}$

$=\{x:\text{for any } S\in\mathcal{W}, \text{ not } (x\in S)\}$

$=\{x:\text{for any } S\in\mathcal{W}, x\in\mathcal{C} S\}$

$=\bigcap_{S\in\mathcal{W}} \mathcal{C} S.$

4.12   *Exercise*

(1) Let $\mathcal{W}$ be the collection $\{A, B, C, D\}$ where $A=\{1, 3, 4\}$, $B=\{1, 2, 4\}$, $C=\{2, 3, 4\}$, $D=\{1, 4\}$. What are the elements of the sets

(i) $\bigcup_{S\in\mathcal{W}} S$ (ii) $\bigcap_{S\in\mathcal{W}} S$?

(2) Let $\mathcal{W}$ be any collection of subsets of a universal set U. Let $A\in\mathcal{W}$ and let B be any other set in U. Prove the following.

(i) $A\subset\bigcup_{S\in\mathcal{W}} S$ (ii) $A\supset\bigcap_{S\in\mathcal{W}} S$

(iii) $B\cup\left\{\bigcap_{S\in\mathcal{W}} S\right\}=\bigcap_{S\in\mathcal{W}} \{B\cup S\}$ (iv) $B\cap\left\{\bigcup_{S\in\mathcal{W}} S\right\}=\bigcup_{S\in\mathcal{W}} \{B\cap S\}$

(v) $\mathcal{C}\left(\bigcap_{S\in\mathcal{W}} S\right)=\bigcup_{S\in\mathcal{W}} \mathcal{C} S$

†(3) Explain why

(i) $\bigcup_{S\in\emptyset} S=\emptyset$ (ii) $\bigcap_{S\in\emptyset} S=U$

4.13†   **Zermelo–Fraenkel set theory**

In Zermelo–Fraenkel set theory the idea is to express all of mathematics in the language of set theory. The price one has to pay for this is that all mathematical objects have to be defined as sets. The reward is that all mathematical assertions can be expressed in terms of a single predicate

$$x\in y$$

and all proofs referred to a single set of axioms – i.e. the axioms of set theory.

We do not attempt a systematic account of the Zermelo–Fraenkel programme in this book. An attempt to do so would leave room for very little else. In later chapters, however, we have been careful to indicate at each stage how the mathematical objects being introduced *can* be defined as special kinds of sets (although we shall certainly not insist on such an interpretation). Those seeking a more formal account of the theory will find A. Abian's *Theory of Sets and Transfinite Arithmetic* (Saunders, 1965) a useful reference.

In this section, we propose to make some remarks about the basic set-theoretic assumptions on which the Zermelo–Fraenkel programme is based. The nature of these assumptions is such that it is natural to take them completely for granted without a second thought and this will be our attitude to the assumptions in the main body of the text. This section is therefore one which can be skipped without remorse if its content seems dull or difficult.

The basic assumption of the Zermelo–Fraenkel theory is that 'every property defines a set'. One has to be a little careful over the formalisation of this idea because of Russell's paradox. Suppose that we assume the existence of the set y consisting of all sets x which do not belong to themselves – i.e.

$$y = \{x : x \notin x\}.$$

If $y \in y$, then y must satisfy the defining property of y – i.e. $y \notin y$. On the other hand, if $y \notin y$, then y satisfies the defining property of y and hence $y \in y$. In either case, a contradiction is obtained. The popular version of this paradox concerns a certain barber who shaves everyone in his town who does not shave himself. The question is then: who shaves the barber? All possible answers lead to contradictions and one concludes that there is no such barber. Similarly, in Zermelo–Fraenkel set theory one concludes that it is meaningless to speak of 'the set of all sets which do not belong to themselves'. It is also meaningless to speak of 'the sets of all sets'. There is no such set.

The assumption that asserts that 'properties define sets' is called the *axiom of replacement*. For our purposes, the following version is adequate. Suppose that u is a set and that $P(x)$ is a predicate. Then

$$\{x : x \in u \text{ and } P(x)\}$$

is a set. Here, of course, u plays the role of the universal set U introduced in the previous chapter and the fact that there is no 'set of all sets' explains our insistence that U will depend on the context.

Next one needs an assumption that asserts that, if x is a set, then the union and intersection of all sets in x is also a set. This assumption is called the *axiom of unions*. One also needs an assumption that asserts the existence of the set of all subsets of a given set s. The set of all subsets of s is called the *power set* of s and denoted by $\mathscr{P}(s)$. (The reason for this nomenclature is that, if s contains n elements, then $\mathscr{P}(s)$ contains 2^n elements.) The axiom which asserts the existence of $\mathscr{P}(s)$ is therefore called the *power set axiom*.

In Zermelo–Fraenkel theory, even the idea of 'equals' is not taken for granted. Since all mathematical objects are to be regarded as sets, we only need a definition of what it means to say that two sets are equal. The obvious thing to do is to define two sets to be equal if and only if they have the same elements – i.e. $x = y \Leftrightarrow \forall z (z \in x \Leftrightarrow z \in y)$. Another way of saying this is that two sets are equal if and only if each is a subset of the other. To make the definition useful, another assumption is required. This is called the *axiom of extensionality*. Essentially, it asserts that, if $x = y$, then y can be substituted for x in any statement without altering the truth value of the statement.

Zermelo–Fraenkel theory requires two further assumptions which it is convenient to postpone to later chapters. These are called the *axiom of infinity* and the *axiom of choice*. It is a remarkable fact that the whole of traditional mathematics can be based on the principles of logic and these six simple assumptions about the properties of sets.

5 RELATIONS

5.1 Ordered pairs

The use of an ordered pair (a, b) of real numbers to label a point in the plane will be familiar from co-ordinate geometry.

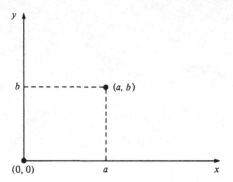

We say that (a, b) is an *ordered* pair because the order in which a and b appear is relevant. For example, (3, 4) does not represent the same point in the plane as (4, 3).

How can we define an ordered pair in general? We certainly do *not* wish to identify (a, b) with the set $\{a, b\}$. Both $\{3, 4\}$ and $\{4, 3\}$, for example, denote the *same* set. For this reason, the set $\{a, b\}$ is often called an *unordered* pair.

A viable alternative is to define

$$(a, b) = \{a, \{a, b\}\}$$

– i.e. (a, b) is defined to be the set whose elements are a and the set $\{a, b\}$. However, the precise form of the definition is irrelevant for our purposes. Any definition which ensures that $(a, b) = (c, d)$ if and only if $a = c$ and $b = d$ will do just as well.

5.2 Cartesian products

The *Cartesian product* of two sets A and B is defined by

$$A \times B = \{(a, b) : a \in A \text{ and } b \in B\}.$$

The notation $A \times B$ is used because, for example, if A has two elements and B has three elements, then $A \times B$ has $6 = 2 \times 3$ elements.

When A and B are sets of real numbers, we can represent $A \times B$ as a set of points in the plane.

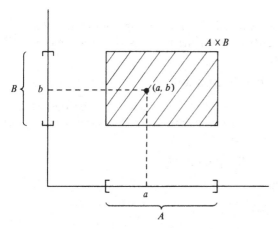

We use the abbreviation $A \times A = A^2$. Thus $\mathbb{R} \times \mathbb{R} = \mathbb{R}^2$ represents the whole plane.

In a similar way, one can introduce ordered n-tuples $(a_1, a_2, \ldots, a_n)$ and consider the Cartesian product of n sets. In particular,

$$\mathbb{R}^n = \{(x_1, x_2, \ldots, x_n) : x_1 \in \mathbb{R}, \ x_2 \in \mathbb{R}, \ldots \text{ and } x_n \in \mathbb{R}\}.$$

5.3   Relations

Any subset R of $A \times B$ defines a *relation* between A and B. If $(a, b) \in R$, we say that the relation R holds between a and b and write

$$a \ R \ b.$$

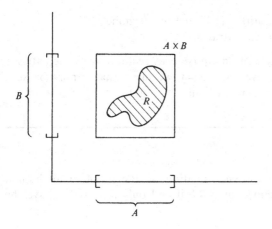

Obviously a relation R is determined if and only if, for each $a \in A$ and $b \in B$, it is known whether or not it is true that $a\,R\,b$.

5.4   Examples

(i) Let A and B both be the set of all human beings and write $a\,R\,b$ if and only if 'a loves b'. In accordance with the remarks above we may identify the relation R with the set

$$R = \{(a,\,b) : a \text{ loves } b\}.$$

(ii) Let A be the set $\mathbb{N}$ of natural numbers and let B be the set $\mathbb{Z}$ of integers. We write $a|b$ if and only if 'a divides b', i.e. $b = ma$ for some integer m. Some ordered pairs in the set

$$R = \{(a,\,b) : a|b\}$$

are $(2,\,6)$, $(3,\,-21)$, $(1,\,7)$.

5.5   Equivalence relations

The idea of equivalence is important not only in mathematics but in life in general. Indeed, all abstractions are based on this idea. A Slovenian peasant, for example, who considers that the only relevant criterion in choosing a wife is that she be sturdily built is splitting the set of available women into two 'equivalence classes'. He has invented an equivalence relation with respect to which all sturdy women are equivalent, regardless of their personal appearance or domestic skills. We have met the same idea in logic. Two statements were said to be equivalent if they had the same truth value, regardless of the subject matter with which they were concerned.

In mathematics, an *equivalence relation* R on a set A is defined to be a relation between the set A and itself which satisfies the following requirements for each a, b and c in A:

(i) $a\,R\,a$ (reflexivity)   (ii) $a\,R\,b \Leftrightarrow b\,R\,a$ (symmetry)
(iii) $a\,R\,b$ and $b\,R\,c \Rightarrow a\,R\,c$ (transitivity).

The most important example of an equivalence relation in mathematics is the relation '*equals*'. When we write $x = y$ we mean that y may be substituted for x in any statement without altering the truth value of the statement. (See §4.13.)

5.6   Examples

(i) An example of an equivalence relation on the set A of all human beings is the relation R defined by $a\,R\,b$ if and only if 'a and b have the same mother'.

(ii) An equivalence relation on the set $\mathbb{Z}$ of integers may be obtained by writing $a \, R \, b$ if and only if a and b have the same remainder when divided by 3.

Let R be an equivalence relation on A. If $a_1 \in A$ write

$$A_1 = \{a : a \, R \, a_1\}.$$

Then A_1 is the set of all $a \in A$ which are equivalent to a_1. We call A_1 an *equivalence class* of R.

As an example, consider logical equivalence on the set of all statements. There are two equivalence classes: the class of true statements and the class of false statements.

5.7   *Theorem* Two distinct equivalence classes are disjoint (i.e. have no points in common).

Proof Let R be an equivalence relation on a set A. If $a_1 \in A$ and $a_2 \in A$, write $A_1 = \{a : a \, R \, a_1\}$ and $A_2 = \{a : a \, R \, a_2\}$.

Suppose that A_1 and A_2 are *not* disjoint. Then they have an element b in common. We shall show that this implies that $A_1 = A_2$ – i.e. A_1 and A_2 are *not* distinct.

Since $b \in A_1$ and $b \in A_2$, $b \, R \, a_1$ and $b \, R \, a_2$. By the symmetric and transitive properties for R it follows that $a_1 \, R \, a_2$.

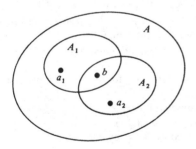

Now suppose that $c \in A_1$. Then $c \, R \, a_1$. Hence $c \, R \, a_2$ by the transitive property. Thus $c \in A_2$. It follows that $A_1 \subset A_2$.

A similar argument shows that $A_2 \subset A_1$ and so we conclude that $A_1 = A_2$.

5.8   Orderings

An *ordering* $\trianglelefteq$ on a set A is a relation between A and itself which satisfies the following properties for each a, b and c in A:

(i) $a \trianglelefteq b$ or $b \trianglelefteq a$ (totality)

(ii) $a \leqslant b$ and $b \leqslant a \Rightarrow a = b$ (antisymmetry)
(iii) $a \leqslant b$ and $b \leqslant c \Rightarrow a \leqslant c$ (transitivity).

If the totality condition is replaced by reflexivity (i.e. $a \leqslant a$), we call $\leqslant$ a *partial* ordering. With a partial ordering, certain pairs of elements in A may not be comparable. When discussing both partial orderings and orderings, we sometimes stress the difference by calling an ordering a *total ordering*.

5.9   *Examples*

(i) The set $\mathbb{R}$ of all real numbers is ordered by the relation $\leqq$ of increasing magnitude. This is the ordering with which we shall be chiefly concerned in this book.

(ii) If A is a set, its power set $\mathscr{P}(A)$ is the set whose elements are the subsets of A. For example, if $A = \{1, 2\}$, then $\mathscr{P}(A) = \{\phi, \{1\}, \{2\}, \{1, 2\}\}$. Observe that the relation $\subset$ is a *partial* ordering on $\mathscr{P}(A)$.

5.10   *Exercise*

(1) Let $A = \{1, 2, 3, 4, 5\}$. Let R be the relation between A and itself defined by $a \, R \, b$ if and only if a and b have the same remainder when divided by 2. Prove that R is an equivalence relation on A and determine the equivalence classes it induces on A.

(2) Write $a|b$ if a and b are natural numbers and a divides b exactly. Show that the relation so defined is a partial ordering on $\mathbb{N}$.

(3) Explain why $\Leftrightarrow$ is an equivalence relation on the set of all statements.

†(4) What is wrong with the following argument which purports to show that the reflexive property for an equivalence relation may be deduced from the other two properties? By the symmetry property, $a \, R \, b$ and $b \, R \, a$. Taking $c = a$ in the transitivity property then yields $a \, R \, a$.

Show, however, that the reflexivity property for an ordering may be deduced from the totality condition.

†(5) Let $\mathscr{P}(A)$ denote the set of all subsets of a set A.
   (i) Prove that $\subset$ is a partial ordering on $\mathscr{P}(A)$.
   (ii) Show that the relation R defined on $\mathscr{P}(A)$ by

$$a \, R \, b \Leftrightarrow \forall x (x \in A \Leftrightarrow x \in B)$$

is an equivalence relation on $\mathscr{P}(A)$.

†(6) Let $\leqslant$ be a total ordering on a set A. Define the associated 'strict ordering' relation $\lhd$ by

$$a \lhd b \Leftrightarrow (a \leqslant b \text{ and } a \neq b).$$

Prove that $\lhd$ is a transitive relation. Is $\lhd$ a total relation? Is $\lhd$ antisymmetric?

6 FUNCTIONS

6.1 Formal definition

A function $f: A \rightarrow B$ is usually said in elementary texts to be a rule which assigns a unique element $y \in B$ to each element $x \in A$. But this is not a very satisfactory definition because it leaves unanswered the question: what is a rule? An answer to this question which is adequate for most elementary applications is that a rule is an algebraic formula. Thus, for example, the formula

$$y = g(x) = x^2 + 1$$

defines a function $g: \mathbb{R} \rightarrow \mathbb{R}$. Any value $x \in \mathbb{R}$ substituted in the right-hand side will yield a unique corresponding value of $y \in \mathbb{R}$ on the left-hand side.

But we shall wish to define functions in more complicated ways than this. The expressions $h(1) = 1$ and

$$h(n+1) = \tfrac{1}{2}\left\{ h(n) + \frac{2}{h(n)} \right\},$$

for example, define a function $h: \mathbb{N} \rightarrow \mathbb{R}$ but one does not calculate $h(n)$ simply by substituting n in an algebraic formula. This is an example of a recursive or inductive definition of a function.

In order to accommodate this and other even more complicated ways of

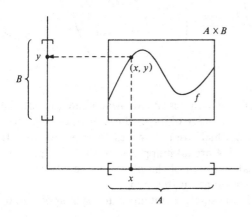

defining a function, we use the following formal definition. If A and B are sets, we say that a *function* $f: A \to B$ is a relation between A and B with the property that, given *any* $x \in A$, there exists a *unique* $y \in B$ such that $(x, y) \in f$. This definition amounts to identifying a function with its *graph*.

The importance of the idea of a function is attested to by the large number of synonyms for the word 'function'. The words '*mapping*', '*operator*' and '*transformation*' are just three of the many alternatives. Which word is used depends on the context. For example, it is sometimes useful to interpret a function $f: A \to B$ as a 'black box' into which an element $x \in A$ is inserted as an input. The 'black box' then does something to x, the result of which is the output $f(x)$.

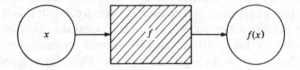

In this context one often calls the function an operator or transformation. Thus one would think of the function $g: \mathbb{R} \to \mathbb{R}$ given above as transforming x into y by squaring it and adding one.

Alternatively, one may think of a function $f: A \to B$ as describing the efforts of a mapmaker who seeks to draw a map of country A on a piece of paper B. It is in this context that one refers to a function as a *mapping*. A point $x \in A$ is said to be *mapped* onto the point $f(x) \in B$. Sometimes $f(x)$ is said to be the *image* of x under the function f.

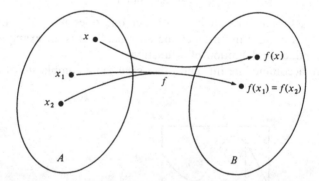

When thinking in these terms, it is necessary to bear in mind that the mapmaker may not use up all of the piece of paper B in drawing the map of A – i.e. there may be points of B which are the images of *no* points of A. It may also be that several points of A are all mapped onto the same point of B. For example, on a large scale map all of the points in the city of Paris might be represented by a single point on the map.

What is *not* admissible is for a single point in A to be mapped onto several points of *B*. Each point $x \in A$ has a *unique* image $y \in B$. One sometimes reads of 'one–many functions' but this is an abuse of language which we shall rigorously avoid.

6.2   Terminology

If $f: A \to B$ is a function from A to B, we call the set A the *domain* of f and the set B the *codomain* of f. If $S \subset A$, we call the set

$$f(S) = \{f(x): x \in S\}$$

the *image* of the set S under the function f. In particular, $f(A)$ is called the *range* of f. Note that the range of a function need not be the whole of the codomain – i.e. it need *not* be true that $f(A) = B$.

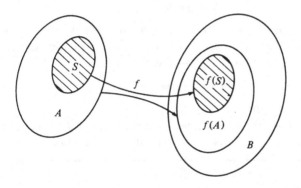

If $T \subset B$, we call the set

$$f^{-1}(T) = \{x : f(x) \in T\}$$

the *pre-image* of the set T under the function f.

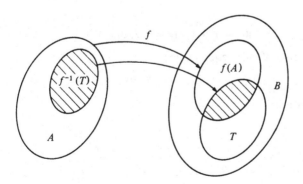

The notation $f^{-1}(T)$ is in some ways an unfortunate one since it seems to imply the existence of an inverse function to f. Note, however, that it makes perfectly good sense to talk about the pre-image of a subset T of B even though an inverse function does *not* exist.

Let $f: A \rightarrow B$ be a function from A to B. It need not be true that $B = f(A)$. However, if it *is* true that $B = f(A)$ we say that f is a function from A *onto* B or that f is a *surjective* function. Thus, f is surjective if and only if *each* element y of the codomain B is the image of an element x of the domain.

Again let $f: A \rightarrow B$. Suppose that, for each $y \in f(A)$, there is a *unique* $x \in A$ such that $y = f(x)$. Then we say that f is a 1:1 function from A to B or that f is an *injective* function. Thus, f is injective if and only if $f(x_1) = f(x_2) \Rightarrow x_1 = x_2$.

If $f: A \rightarrow B$ is a 1:1 function from A onto B (i.e. f is both surjective and injective), then we call f a *bijection*. The requirement that $f: A \rightarrow B$ is a bijection can be rephrased as the assertion that the equation

$$y = f(x)$$

has a *unique* solution $x \in A$ for *each* $y \in B$. But this is precisely what is required in order that the equation $y = f(x)$ define x as a function of y – i.e. that there exists a function $f^{-1}: B \rightarrow A$ such that

$$x = f^{-1}(y) \Leftrightarrow y = f(x)$$

for each $x \in A$ and $y \in B$.

We call $f^{-1}: B \rightarrow A$ the *inverse function* of the function $f: A \rightarrow B$. Note that an inverse function to f exists if and only if f is a bijection.

6.3   Examples

(i) A sequence is simply a function whose domain is the set $\mathbb{N}$ of all natural numbers. The nth term of a sequence $f: \mathbb{N} \rightarrow \mathbb{R}$ is defined to be $x_n = f(n)$. We use the notation $\langle x_n \rangle$ to denote the sequence whose nth term is x_n. Thus the sequence $\langle n^2 + 1 \rangle$ is to be identified with the function $g: \mathbb{N} \rightarrow \mathbb{R}$ defined by

$$g(n) = n^2 + 1 \quad (n = 1, 2, 3, \ldots).$$

The first few terms of this sequence are

$$2, 5, 10, 17, 26, 37, 50, \ldots.$$

(ii) The equations

$$\left.\begin{array}{l} u = x^2 + y^2 \\ v = x^2 - y^2 \end{array}\right\}$$

define a function $f: \mathbb{R}^2 \rightarrow \mathbb{R}^2$. The image of (x, y) under f is $(u, v) = (x^2 + y^2, x^2 - y^2)$.

Observe that f is *not* surjective. If u and v are given by the above equations, then

$$\left.\begin{array}{c} u+v=2x^2\geqq0 \\ u-v=2y^2\geqq0. \end{array}\right\}$$

and hence, for example, the point $(u, v)=(-1, -1)$ is not the image of anything under this function. In fact

$$f(\mathbb{R}^2)=\{(u, v):u\geqq v \text{ and } u\geqq -v\}.$$

Nor is f *injective*. The equations

$$\left.\begin{array}{c} 5=x^2+y^2 \\ 3=x^2-y^2 \end{array}\right\}$$

have *four* solutions for (x, y) – namely $(2, 1)$, $(2, -1)$, $(-2, 1)$ and $(-2, -1)$.

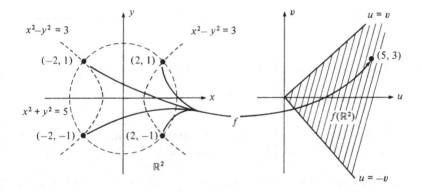

(iii) Let $A=\{x:x\geqq0\}$ and $B=\{y:y\geqq1\}$. The function $f: A\to B$ defined by

$$f(x)=x^2+1 \quad (x\geqq0)$$

is a bijection and hence has an inverse function $f^{-1}: B\to A$. Since, for each $x\geqq0$ and $y\geqq1$,

$$x=f^{-1}(y)\Leftrightarrow y=f(x),$$

a formula for f^{-1} can be obtained by solving the equation $y=x^2+1$ to obtain x in terms of y. Because $y\geqq1$, the equation has a solution. Moreover, as $x\geqq0$, the solution is unique and is given by $x=\sqrt{(y-1)}$. (Remember that, if $z\geqq0$, then $\sqrt{z}$ denotes the *non-negative* number whose square is z.) Thus

$$f^{-1}(y)=\sqrt{(y-1)} \quad (y\geqq1).$$

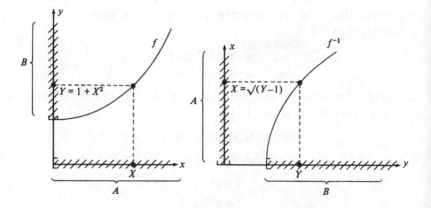

Note that either graph can be obtained from the other by a rotation about the line $x = y$.

6.4   Exercise

(1) Consider the function $f: \mathbb{R}^2 \to \mathbb{R}^2$ defined by $f(x, y) = (u, v)$ where

$$\left. \begin{array}{l} u = x + y \\ v = x - y. \end{array} \right\}$$

Prove that f is a bijection and obtain a formula for its inverse function. Find $f(S)$ and $f^{-1}(T)$ in the case $S = T = \{(x, y) : 0 \leq x \leq 1 \text{ and } 0 \leq y \leq 1\}$.

(2) Let $A = \{\theta : 0 < \theta \leq \tfrac{1}{2}\pi\}$ and $B = \{\phi : -\pi < \phi \leq \pi\}$. A function $g: A \times B \to \mathbb{R}^2$ is defined by $(x, y) = g(\theta, \phi)$ where

$$\left. \begin{array}{l} x = a \sin \theta \cos \phi \\ y = b \sin \theta \sin \phi. \end{array} \right\}$$

Prove that the range of g is the set

$$F = \left\{ (x, y) \mid 0 < \frac{x^2}{a^2} + \frac{y^2}{b^2} \leq 1 \right\}$$

and show that g is injective.

(3) Let $f: A \to B$. If $S_1 \subset S_2 \subset A$, show that $f(S_1) \subset f(S_2) \subset B$. If $T_1 \subset T_2 \subset B$, show that $f^{-1}(T_1) \subset f^{-1}(T_2) \subset A$.

(4) Let X denote a collection of subsets of A and Y a collection of subsets of B. If $f: A \to B$, prove that

(i) $f\left(\bigcup_{S \in X} S \right) = \bigcup_{S \in X} f(S)$ (ii) $f\left(\bigcap_{S \in X} S \right) \subset \bigcap_{S \in X} f(S)$

(iii) $f^{-1}\left(\bigcup_{T \in Y} T\right) = \bigcup_{T \in Y} f^{-1}(T)$ $f^{-1}\left(\bigcap_{T \in Y} T\right) = \bigcap_{T \in Y} f^{-1}(T)$.

Give an example to show that equality need *not* hold in (ii).

(5) Let $f: A \to B$.

(i) Prove that, for each $S \subset A$,

$$f^{-1}(f(S)) \supset S.$$

Show that f is injective if and only if $f^{-1}(f(S)) = S$ for each $S \subset A$.

(ii) Prove that, for each $T \subset B$,

$$f(f^{-1}(T)) \subset T.$$

Show that, for each $T \subset f(A)$, $f(f^{-1}(T)) = T$. Deduce that f is surjective if and only if $f(f^{-1}(T)) = T$ for each $T \subset B$.

(6) Suppose that $A_1 \subset A_2$ and consider two functions $f_1: A_1 \to B$ and $f_2: A_2 \to B$ which satisfy

$$f_1(x) = f_2(x)$$

for each $x \in A_1$. We say that f_1 is the *restriction* of f_2 to A_1 and that f_2 is an *extension* of f_1 to A_2.

Define $f: \mathbb{R} \to \mathbb{R}$ by $f(x) = |x|$. Let g be the restriction of f to $(0, \infty)$. Find an extension of g to $\mathbb{R}$ which is differentiable at every point of $\mathbb{R}$.

6.5   Composition

Suppose that $f: B \to C$ and $g: A \to B$. Then their *composition* $f \circ g : A \to C$ is defined by

$$f \circ g(x) = f(g(x)) \quad (x \in A).$$

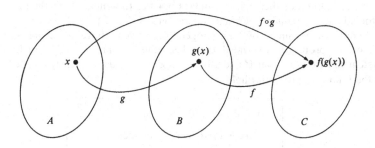

Suppose that $f: A \to B$ is a bijection. Then it has an inverse function $f^{-1}: B \to A$. The composite functions $f \circ f^{-1}: B \to B$ and $f^{-1} \circ f: A \to A$ are then both *identity* functions – i.e. they map each point onto itself. To see this, recall that

$$y=f(x) \Leftrightarrow x=f^{-1}(y)$$

and hence

$$f \circ f^{-1}(y)=f(f^{-1}(y))=f(x)=y \quad (y \in B)$$
$$f^{-1} \circ f(x)=f^{-1}(f(x))=f^{-1}(y)=x \quad (x \in A).$$

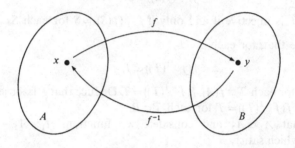

6.6†   Binary operations and groups

Suppose that $\mathcal{G}$ is a set and that ϕ is a function from $\mathcal{G} \times \mathcal{G}$ to $\mathcal{G}$. Such a function ϕ is often called a '*binary operation*' on $\mathcal{G}$ because *two* elements of $\mathcal{G}$ are put together to make a new element of $\mathcal{G}$.

Addition and multiplication can be regarded as binary operations on the set $\mathbb{R}$ of real numbers with $+:\mathbb{R} \times \mathbb{R} \to \mathbb{R}$ and $\times:\mathbb{R} \times \mathbb{R} \to \mathbb{R}$ defined by

$$+(a, b)=a+b$$
$$\times(a, b)=ab.$$

Similarly, if A is any set, we can think of composition as a binary operation defined on the set $\mathcal{A}$ of all functions $f: A \to A$. Then $\circ: \mathcal{A} \times \mathcal{A} \to \mathcal{A}$ is defined by

$$\circ(f, g)=f \circ g.$$

As these examples indicate, it is standard practice to write $a * b$ for the image of the ordered pair (a, b) under the binary operation $*$.

The three examples of binary operations given above all share some very important properties (provided that their domains are suitably restricted). The system consisting of a set $\mathcal{G}$ and a binary operation $*$ defined on $\mathcal{G}$ is called a *group* if the following properties hold.

> (i) For any a, b and c in $\mathcal{G}$,
>
> $a*(b*c)=(a*b)*c$ (associative law).
>
> (ii) There exists a unique $e \in \mathcal{G}$ such
>   that for any $a \in \mathcal{G}$,
>
> $a*e=e*a=a$ (law of the unit).
>
> (iii) For any $a \in \mathcal{G}$, there exists a
>   unique $b \in \mathcal{G}$ such that
>
> $a*b=b*a=e$ (law of the inverse).

The real number system ℝ is a group under addition. The element e of item (ii) is zero and the element b of item (iii) is $-a$. The system $ℝ \setminus \{0\}$ (i.e. ℝ with 0 removed) is a group under multiplication. In this case $e = 1$ and $b = 1/a$. Finally, the system $\mathscr{P}$ of all *bijective* functions $f: ℝ \to ℝ$ is a group under composition. Here e is the identity function on ℝ and b is the inverse function to a.

The requirements for a *commutative group* are obtained by adding the following item to the list of group properties.

> (iv) For any a and b in $\mathcal{G}$,
>
> $a*b = b*a$ (commutative law).

The real number system ℝ is thus a commutative group under addition and $ℝ \setminus \{0\}$ is a commutative group under multiplication. However, $\mathscr{P}$ is *not* a commutative group under composition as the following example shows.

6.7   *Example* Define the bijective functions $f: ℝ \to ℝ$ and $g: ℝ \to ℝ$ by

$$f(x) = x^3 \quad (x \in ℝ)$$
$$g(x) = 1 + x \quad (x \in ℝ).$$

Then

$$f \mathbf{O} g(x) = f(g(x)) = (1 + x)^3$$
$$g \mathbf{O} f(x) = g(f(x)) = 1 + x^3.$$

Thus it is *false* that $f \mathbf{O} g = g \mathbf{O} f$.

6.8† Axiom of choice

In §4.13, we discussed very briefly four of the six basic assumptions of the formal theory of sets. In this section, we continue the discussion by introducing a fifth assumption which is called the *axiom of choice*. This asserts that, given any collection X of non-empty sets, there is a way of choosing an element from each of the sets. More precisely, the axiom of choice asserts that there exists a function

$$f: X \to \bigcup_{S \in X} S$$

such that, for each $S \in X$, $f(S) \in S$. One thinks of $f(S)$ as being the element chosen from the set S by the function f.

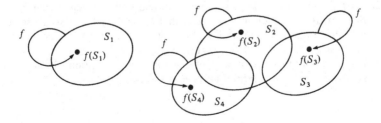

The axiom of choice seems an innocent enough assumption but it is as well to be aware that it has some curious consequences. For example, using the axiom of choice, it is possible to partition the surface of a sphere into three disjoint, congruent sets *A*, *B* and *C* (congruent meaning that any of the sets can be rotated so as to coincide with any of the others). So far this is quite unremarkable. But the sets have the further property that *A* is *also* congruent to $B \cup C$. We therefore have the mathematical equivalent of the miracle of the loaves and fishes. The sphere is carved up into three distinct 'identical' parts but when two of the parts are put together their union is found to be 'identical' to the remaining part. For these sets the notion of 'area' simply makes no sense. We say that the sets are *non-measurable*.

6.9   *Example* A surjective function $f\colon A \to B$ has a *right-inverse*, i.e. there exists a function $g\colon B \to A$ such that

$$f \circ g(y) = y \quad (y \in B).$$

Proof Let X be the collection of all the sets $f^{-1}(\{y\})$ where $y \in B$. Since f is surjective, these sets are all non-empty and hence, by the axiom of choice, there exists a function $h\colon X \to A$ such that $h(f^{-1}\{y\}) \in f^{-1}\{y\}$. Define the function $g\colon B \to A$ by

$$g(y) = h(f^{-1}\{y\}).$$

Since $g(y) \in f^{-1}\{y\}$, $f(g(y)) = y$, i.e. $f \circ g(y) = y$.

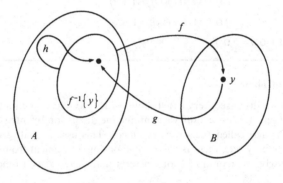

6.10†   *Exercise*

(1) Let $f\colon \mathbb{R} \to \mathbb{R}$ be defined by $f(x) = 1 + x^2$ and let $g\colon \mathbb{R} \to \mathbb{R}$ be defined by $g(x) = x^2$. Prove that $f \circ g \neq g \circ f$.

(2) Let $f\colon A \to B$. Prove that f is surjective if and only if it has a right-inverse, i.e. there exists a function $g\colon B \to A$ such that $f \circ g(y) = y$ for all $y \in B$ (see example 6.9). Prove that f is injective if and only if it has a *left-inverse*, i.e. there exists a function $h\colon B \to A$ such that $h \circ f(x) = x$ *for all* $x \in A$.

(3) Let $f\colon A \to B$. Prove that any of the following conditions is necessary and sufficient for the existence of an inverse function.

    (i) f has a right-inverse and left-inverse

    (ii) f has a unique right-inverse

    (iii) f has a unique left-inverse.

(4) Let $\mathcal{P}(A)$ be the collection of all subsets of A and let $\mathcal{P}(B)$ be the collection of all subsets of B. If $f: A \to B$, we may define a function $F: \mathcal{P}(A) \to \mathcal{P}(B)$ by

$$F(S) = f(S).$$

Does F have an inverse function?

(5) Explain why 'and' can be regarded as a binary operation acting on the set of statements.

(6) A system consisting of a set $\mathcal{G}$ and a binary operation $*$ on $\mathcal{G}$ satisfies

    (i) For any a, b and c in $\mathcal{G}$

    $a*(b*c) = (a*b)*c$ (associative law)

    (ii) there exists an $e \in \mathcal{G}$ such that for any $a \in \mathcal{G}$
    $e*a = a$ (left-identity law)

    (iii) for any $a \in \mathcal{G}$ there exists $b \in \mathcal{G}$ such that
    $b*a = e$ (left-inverse law).

Show that $\mathcal{G}$ is a group under $*$.

7 REAL NUMBERS (I)

7.1 Introduction

In the examples used to illustrate the ideas introduced in previous chapters, the properties of the various number systems have been taken for granted. In the next few chapters, however, we propose to give a *rigorous* account of the theory of the real number system $\mathbb{R}$ and of its subsystems. Such an account is essential if the principles of analysis are to be placed on a firm footing.

The first and perhaps the most difficult step is to try and forget for the moment everything that we already know about number systems. Those with painful memories of schoolroom arithmetic may think this easy enough, but it is a mistake to underestimate the extent of our childhood conditioning about number. In what follows, we propose to take *nothing* about the number systems for granted – not even such statements as

$$2 + 2 = 4.$$

We shall be asking: what does this statement mean? How is it proved? What are the assumptions on which the proof is based? These considerations become meaningless if we cannot free ourselves from the notion that $2 + 2 = 4$ is beyond question.

There are two possible approaches to the study of the number systems. The first is axiomatic and the second is constructive. The two approaches are complementary and we shall consider both.

In this chapter we begin the study of the axiomatic approach. The idea is to give a list of axioms from which all the properties of the real number system may be deduced. It is clearly desirable to keep the number of axioms to a minimum and to make each axiom as simple as possible. Our list of axioms will then provide a concise and easily remembered summary of the logical structure of the real number system.

7.2 Real numbers and length

A mathematical theory begins with a list of axioms. From these axioms the theorems are to be deduced using only the rules of logic. The axioms we shall give for the real number system fall under three headings:

(I) axioms of arithmetic
(II) axioms of order
(III) axiom of the continuum.

The current chapter deals with the axioms of arithmetic and order. The next chapter explains how the properties of the natural numbers (1, 2, 3,...) and other subsystems of ℝ may be deduced. The axiom of the continuum is studied in chapter 9.

On seeing the axioms, one might reasonably ask: why should the theorems deduced from these axioms be of any interest? Or, in other words: what situations within my experience do these axioms fit? These are, of course, questions about the application of the theory rather than the theory itself.

The answer is that the real numbers were introduced originally as an idealisation of the notion of *length*. With this application in mind, it is useful to picture the real numbers as the points along a line which extends indefinitely in both directions. The line may then be regarded as an ideal ruler with which we may measure the lengths of line segments in Euclidean geometry.

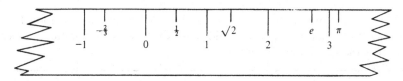

What things should we be able to do with lengths? The idea of *addition* arises from the need to be able to work out the total length of two adjacent line segments. Similarly, *multiplication* arises from the need to be able to work out the area of a rectangle.

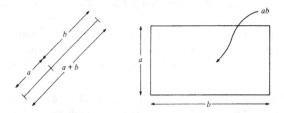

If addition and multiplication are to be interpreted in this way, then they must satisfy certain laws. Otherwise we might get different values for lengths and areas depending on how we went about computing them. The diagrams below, for example, illustrate two ways in which the area A of a rectangle might be evaluated.

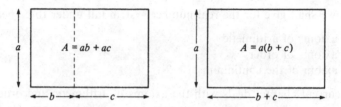

From these diagrams, we are led to propose the rule

$$ab + ac = a(b + c).$$

In the same way, we will be led to other rules concerning addition and multiplication (and subtraction and division). These rules, taken together, are the laws of arithmetic with which we have been familiar since early childhood.

Our first task is to write down a list of axioms which embody these laws of arithmetic. The only surprising thing about these axioms will be that we need so few of them.

Not all the laws of arithmetic are axioms. Those which are not must be deduced from those which are. We shall do this for some of the tricky cases and leave the other cases as exercises.

Those with some training in formal abstract algebra will already have met much of what follows and could well skip forward to chapter 9 which concerns the continuum axiom.

7.3   Axioms of arithmetic

We begin by assuming that $\mathbb{R}$ is a set and that we are given two functions from $\mathbb{R} \times \mathbb{R}$ to $\mathbb{R}$. These functions are called *addition* and *multiplication*. The image of a pair (a, b) of elements of $\mathbb{R}$ under the addition function is denoted by $a + b$. Its image under the multiplication function is denoted by ab.

As explained in §7.2, it is necessary to insist that addition and multiplication satisfy the usual laws of arithmetic. These laws are summarised by the following list of axioms which we require to be satisfied for each a, b and c in the set $\mathbb{R}$.

Algebraists summarise these axioms by saying that a system which satisfies them is a *field*. Thus a field is a system in which the ordinary laws of arithmetic hold. The fact that the ordinary laws of arithmetic hold does not, however, prevent some fields from being rather peculiar (see exercise 7.5(5)).

Note that the requirements for $\mathscr{F}$ to be a field can be neatly summarised

Axiom I	*Axiom II*
(i) $a+(b+c)=(a+b)+c$.	(i) $a(bc)=(ab)c$.
(ii) $a+b=b+a$.	(ii) $ab=ba$.
(iii) There exists a unique element $0\in\mathbb{R}$ such that $a+0=a$.	(iii) There exists a unique element $1\in\mathbb{R}$ $(1\neq0)$ such that $a1=a$.
(iv) For any $a\in\mathbb{R}$ there exists a unique $x\in\mathbb{R}$ such that $a+x=0$.	(iv) For any $a\in\mathbb{R}$ (*except* $a=0$) there exists a unique $y\in\mathbb{R}$ such that $ay=1$.

Axiom III
$a(b+c)=ab+ac$.

in terms of the notion of a *group* (§6.6) For $\mathcal{F}$ to be a field, we require that

(i) $\mathcal{F}$ is a commutative group under addition.
(ii) $\mathcal{F}\setminus\{0\}$ is a commutative group under multiplication.
(iii) Addition and multiplication are linked by the distributive law (i.e. axiom III).

Apart from the axioms, we need a number of abbreviations and conventions. The first of these arises from axiom Ii. This asserts that it does not matter how brackets are inserted when adding three quantities. Thus, we might just as well write $a+b+c$ instead of $a+(b+c)$ or $(a+b)+c$. Similarly, we write abc instead of $a(bc)$ or $(ab)c$.

The element x of axiom Iiv is usually denoted by $(-a)$ and the element y of axiom IIiv by a^{-1} (provided $a\neq0$). Thus $a+(-a)=0$ and $aa^{-1}=1$.

Subtraction and *division* are defined by

$$b-a=b+(-a)$$

$$\frac{b}{a}=ba^{-1} \quad \text{(provided that } a\neq0\text{)}.$$

Division by zero is excluded by axiom IIiv. Thus the expression $x/0$ makes no sense at all. In particular, $x/0$ is *not* 'equal to ∞'. We shall have a great deal of use for the symbol '∞' but it must be clearly understood that ∞ is *not* a real number.

We use the symbol 2 to stand for $1+1$ and the symbol 3 to stand for $2+1$ and so on. Thus, for example,

$$4 = 3+1 \quad \text{(definition)}$$
$$= (2+1)+1 \quad \text{(definition)}$$
$$= 2+(1+1) \quad \text{(axiom Ii)}$$
$$= 2+2 \quad \text{(definition)}.$$

7.4   *Example*

$x+x+x = 3x.$

Proof
$$x+x+x = (x+x)+x \quad \text{(definition)}$$
$$= x(1+1)+x \quad \text{(axioms IIiii and III)}$$
$$= 2x+x \quad \text{(definition)}$$
$$= x(2+1) \quad \text{(axiom III)}$$
$$= x3 \quad \text{(definition)}$$
$$= 3x \quad \text{(axiom IIii)}.$$

7.5   *Exercise*

(1) Prove that, for all a, b and c in the set $\mathbb{R}$,

(i) $(a+b)-c = a+(b-c)$
(ii) $(ab)/c = a(b/c)$ (provided $c \neq 0$).

(2) If x^2 stands for xx, prove that, for all a and b in the set $\mathbb{R}$,

$$(a+b)^2 = a^2 + 2ab + b^2.$$

(3) Prove that, for all a and b in the set $\mathbb{R}$,

(i) $a+b = a+c$ implies $b = c$.
(ii) If $a \neq 0$, then $ab = ac$ implies $b = c$.

†(4) Let A be a set. Which of the axioms of arithmetic hold if a, b and c are subsets of A and $a+b$ is interpreted as $a \cup b$ and ab is interpreted as $a \cap b$?

†(5) Let $\mathscr{F} = \{0, 1\}$ and suppose that addition and multiplication are defined on $\mathscr{F}$ *not* in the usual way but by means of the 'peculiar' addition and multiplication tables below:

+	0	1
0	0	1
1	1	0

×	0	1
0	0	0
1	0	1

(Note that $1+1=0$.) Show, by considering all possible values for a, b and c, that $\mathcal{F}$ is a field under the 'peculiar' rules of addition and multiplication given.
†(6) Suppose that $\mathcal{F}$ is a field under $+$ and $\times$. Let $\mathcal{H}$ be a subset of $\mathcal{F}$ containing at least two elements. Show that $\mathcal{H}$ is a field under $+$ and $\times$ if and only if, for all non-zero elements a and b in $\mathcal{H}$, $a-b$ and a/b are also elements of $\mathcal{H}$.

If $\mathcal{G}$ is a group under $*$ and $\mathcal{H} \subset \mathcal{G}$, show that $\mathcal{H}$ is a group under $*$ if and only if, for all elements a and b in $\mathcal{H}$, $a*b^{-1}$ is also an element of $\mathcal{H}$ (where b^{-1} is the inverse element to b).

7.6 Some theorems in arithmetic

7.7 *Theorem* For any $a \in \mathbb{R}$,
$$a0 = 0.$$

Proof We have that
$$a+0=a \quad \text{(axiom Iiii)}$$
$$a(a+0)=aa$$
$$aa+a0=aa \quad \text{(axiom III)}$$
$$aa+a0=aa+0 \quad \text{(axiom Iiii)}$$
$$a0=0 \quad \text{(exercise 7.5(3i))}.$$

[Since $a+0=a$, the second step may be obtained by substituting $a+0$ for the second a which appears on the left-hand side of the equation $aa=aa$ (see §5.5).]

7.8 *Theorem* For any $a \in \mathbb{R}$ and any $b \in \mathbb{R}$,
$$a(-b) = -(ab) \quad \text{(i.e. plus} \times \text{minus} = \text{minus)}$$

Proof We have that
$$b+(-b)=0 \quad \text{(axiom Iiv)}$$
$$a(b+(-b))=a0=0 \quad \text{(theorem 7.7)}$$
$$ab+a(-b)=0 \quad \text{(axiom III)}$$
$$ab+a(-b)=ab+(-(ab)) \quad \text{(axiom Iiv)}$$
$$a(-b)=-(ab) \quad \text{(exercise 7.5(3i))}.$$

7.9 *Exercise*
(1) Prove that, for any $a \in \mathbb{R}$ and any $b \in \mathbb{R}$,
$$(-a)(-b)=ab \quad \text{(i.e. minus} \times \text{minus} = \text{plus)}.$$

(2) Prove that, if $a \in \mathbb{R}$ and $a \neq 0$, then

$$(-a)^{-1} = -(a^{-1}).$$

(3) Suppose that $a \in \mathbb{R}$ and $b \in \mathbb{R}$. Prove that

$$ab = 0 \Rightarrow a = 0 \text{ or } b = 0.$$

(4) Prove that for any $a \in \mathbb{R}$ and any $b \in \mathbb{R}$,

$$a^2 - b^2 = (a - b)(a + b).$$

7.10   Axioms of order

The general properties of an ordering were introduced in chapter 5 but we now have a further factor to take into account. If we are to introduce an ordering into the real number system, this ordering must be consistent with the arithmetic of the real number system.

For example, of two rectangles with the same width, we want the rectangle of greater length to have the greater area. Thus, if $c > 0$ and $a > b$, we want it to be true that $ac > bc$.

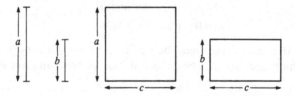

One way of introducing an ordering which is consistent with the arithmetic into the system is by splitting the non-zero elements of $\mathbb{R}$ into two disjoint sets, $\mathbb{R}_+$ (the *positive* elements) and $\mathbb{R}_-$ (the *negative* elements), in such a way that

(a) $x \in \mathbb{R}_+ \Leftrightarrow -x \in \mathbb{R}_-$
(b) If $a \in \mathbb{R}_+$ and $b \in \mathbb{R}_+$, then $a + b \in \mathbb{R}_+$ and $ab \in \mathbb{R}_+$.

If we write '$x > 0$' instead of 'x is positive', these requirements are embodied in the list given below.

Axiom IV

(i) For each $x \in \mathbb{R}$, one and only one of
$x > 0$, $x = 0$, $-x > 0$ is true.
(ii) If $a > 0$ and $b > 0$, then $a + b > 0$.
(iii) If $a > 0$ and $b > 0$, then $ab > 0$.

Given axiom IV, we define

$$a > b$$

to mean that $a - b > 0$. The expression $a < b$ is defined to mean the same as $b > a$.

It is also useful to define $a \geq b$. This means '$a > b$ or $a = b$'. Similarly, $a \leq b$ means '$a < b$ or $a = b$'.

7.11   *Theorem* $1 > 0$.

Proof Suppose it is false that $1 > 0$. Since $1 \neq 0$ (axiom IIiii), it follows that

$$-1 > 0 \quad \text{(axiom IVi)}$$
$$(-1)(-1) > 0 \quad \text{(axiom IViii)}$$
$$1 > 0 \quad \text{(exercise 7.9(1))}.$$

But $-1 > 0$ and $1 > 0$ cannot both be true (axiom IVi) and hence we have a contradiction. Thus $1 > 0$. [See exercise 2.11(6).]

7.12   *Theorem* The relation $\leq$ is an ordering on $\mathbb{R}$.

Proof We have to check the requirements for an ordering given in §5.8:

(i) $a \leq b$ or $b \leq a$
(ii) $a \leq b$ and $b \leq a \Rightarrow a = b$
(iii) $a \leq b$ and $b \leq c \Rightarrow a \leq c$.

The first two requirements follow immediately from axiom IVi and so we consider only the third.

If $a = b$ or $b = c$, there is nothing to prove. If $a < b$ and $b < c$, we have that

$$b - a > 0 \quad \text{(definition)}$$
$$c - b > 0 \quad \text{(definition)}$$
$$c - a = (c - b) + (b - a) > 0 \quad \text{(axiom IVii)}$$
$$a < c \quad \text{(definition)}.$$

7.13   **Intervals**

An *interval* I is a subset of $\mathbb{R}$ with the property that, if $y \in I$ and $z \in I$, then

$$y < x < z \Rightarrow x \in I.$$

Thus I contains all the real numbers between any pair of its elements. The following examples are said to be *open intervals*:

$$(a, b) = \{x : a < x < b\}$$
$$(a, \infty) = \{x : a < x\}$$
$$(-\infty, b) = \{x : x < b\}.$$

The intervals

$$[a, b] = \{x : a \leq x \leq b\}$$
$$[a, \infty) = \{x : a \leq x\}$$
$$(-\infty, b] = \{x : x \geq b\}$$

are said to be *closed*. The sets $\emptyset$ and $\mathbb{R}$ are also intervals and these are allowed to count both as open and as closed intervals. An example of an interval which is neither open nor closed is $[a, b) = \{x : a \leq x < b\}$.

Some comments on the notation may be helpful. There is some risk of confusing the open interval (a, b) with the ordered pair (a, b) since the same notation is used for both. Some authors use $]a, b[$ instead of (a, b) for this reason. There is also the question of the usage of the symbols ∞ and $-\infty$. Recall from §7.3 that these do *not* represent real numbers and their use in this context is simply to provide a picturesque and easily remembered notation. For example, the 'endpoints' of the interval $[a, b)$ are the real numbers a and b. But the interval $(-\infty, b)$ has only *one* endpoint namely b.

7.14   *Example* Find the set of values of x for which $x^2 - 3x + 2 > 0$.

Since $x^2 - 3x + 2 = (x - 1)(x - 2)$, we have that $(x - 1)(x - 2) > 0$ if and only if

$$\{(x - 1) > 0 \text{ and } (x - 2) > 0\} \quad \text{or} \quad \{(x - 1) < 0 \text{ and } (x - 2) < 0\}$$
$$\{x > 1 \text{ and } x > 2\} \quad \text{or} \quad \{x < 1 \text{ and } x < 2\}$$
$$(x > 2) \quad \text{or} \quad (x < 1).$$

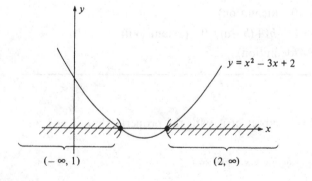

The set required is therefore $(2, \infty) \cup (-\infty, 1)$.

7.15   *Exercise*

(1) Let a, b, c and d be real numbers.

   (i) If $a>b$ and $c>d$, prove that $a+c>b+d$ (i.e inequalities can be added). What conclusion may be drawn from $a\geq b$ and $c\geq d$; and what conclusion from $a>b$ and $c\geq d$?
   (ii) If $c>0$ and $a>b$, prove that $ac>bc$ (i.e. inequalities can be multiplied through by a positive factor). What conclusion may be drawn from $c>0$ and $a\geq b$?
   (iii) If $c<0$ and $a>b$, prove that $ac<bc$ (i.e. multiplication by a *negative* factor *reverses* the direction of an inequality).

(2) Using the knowledge we have so far about inequalities between real numbers, prove that

   (i) $2>1$ (ii) $3>0$ (iii) $2\geq 2$
   (iv) $-1<0$ (v) $-3<-2$ (vi) $2\leq 3$.

(3) In question (1i) we saw that inequalities can be added. Using as examples the inequalities obtained in question 2, show that inequalities *cannot*, in general, be subtracted, multiplied or divided. Show, however, that inequalities between *positive* real numbers can be multiplied.
(4) Suppose that $x>0$ and $y>0$. Prove that $x<y$ if and only if $x^2<y^2$.
(5) For each of the following inequalities, express the set of x for which it is true in terms of intervals (as in example 7.14). Illustrate your results with graphs.

   (i) $x^2+x-2<0$ (ii) $\dfrac{x-1}{x-2}>0$ $(x\neq -2)$.

†(6) Explain why *no* ordering of the field $\mathscr{F}$ of exercise 7.5(5) is compatible with its arithmetic (i.e. all possible orderings violate one of the axioms of order).

8 PRINCIPLE OF INDUCTION

8.1   Ordered fields

A system which satisfies the axioms of arithmetic is called a *field*. A system which satisfies both the axioms of arithmetic and the axioms of order is called an *ordered field*. The real number system ℝ is an ordered field which satisfies a further axiom called the continuum axiom. This is discussed in the next chapter. For the moment we only wish to comment on the fact that, while there are many distinct ordered fields, the system ℝ is unique (See §9.21.)

In this chapter our aim is to introduce the ordered field ℚ of all *rational numbers* (or fractions) and to discuss the reasons why this ordered field is not adequate for the purposes for which we require the real number system ℝ. As a preliminary to this objective, it is necessary to begin by providing a *precise*, formal definition of the system ℕ of *natural numbers* (or whole numbers) and to investigate the properties of this system. The main tool in this investigation is the exceedingly important *principle of induction*.

The sets ℕ and ℚ (and also the set ℤ of integers) will be defined as subsets of ℝ. Note, however, that, since no use at all will be made of the continuum axiom in this chapter, it follows that *any* ordered field contains subsets with the same structure as ℕ, ℚ and ℤ. In particular, any ordered field contains an ordered subfield with the same structure as ℚ. In some sense therefore, ℚ is the 'simplest' possible ordered field.

8.2   The natural numbers

We have defined the numbers 1, 2, 3,... to be 1, $1+1$, $1+1+1$,.... These numbers are called whole numbers or natural numbers. We now need a precise definition of the *set* ℕ of natural numbers. One may well ask: what is wrong with

$$\{1, 2, 3, \ldots\}$$

as a definition for ℕ? The answer lies in the use of the notation '...'. We may feel that we have a good intuitive grasp of what this means but, before we can get down to work, this intuitive understanding has to be put into precise terms.

We proceed as follows. Let $\mathcal{W}$ denote the collection of all sets S of positive real numbers which have the properties

(a) $1 \in S$
(b) $x \in S \Rightarrow x + 1 \in S$.

The collection $\mathcal{W}$ is not empty because it contains the set $(0, \infty)$ of all positive real numbers. We now define the set $\mathbb{N}$ of natural numbers by

$$\mathbb{N} = \bigcap_{S \in \mathcal{W}} S.$$

If $\mathcal{W}$ contained only three sets S_1, S_2 and S_3 (which is certainly not the case) the definition would be illustrated by the accompanying Venn diagram

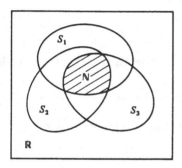

8.3 Theorem

(i) $1 \in \mathbb{N}$.
(ii) $n \in \mathbb{N} \Rightarrow n + 1 \in \mathbb{N}$.

Proof We know that $x \in \mathbb{N}$ if and only if $x \in S$ for any $S \in \mathcal{W}$.
(i) By definition of $\mathcal{W}$, $1 \in S$ for any $S \in \mathcal{W}$. Hence $1 \in \mathbb{N}$.
(ii) Suppose that $n \in \mathbb{N}$. Then $n \in S$ for any $S \in \mathcal{W}$. By definition of $\mathcal{W}$, it follows that $n + 1 \in S$ for any $S \in \mathcal{W}$. Hence $n + 1 \in \mathbb{N}$. This shows that $n \in \mathbb{N}$ implies $n + 1 \in \mathbb{N}$.

Theorem 8.3 says that $\mathbb{N}$ is itself a set in the collection $\mathcal{W}$. Thus $1 \in \mathbb{N}$. But $1 \in \mathbb{N}$ implies $1 + 1 \in \mathbb{N}$. Hence $2 \in \mathbb{N}$. But $2 \in \mathbb{N}$ implies $2 + 1 \in \mathbb{N}$. Proceeding in this way, we can show that any specific whole number is an element of $\mathbb{N}$. Moreover, $\mathbb{N}$ is the *smallest* set in $\mathcal{W}$ because it is contained in every other set in $\mathcal{W}$. This is the basis for our intuition that $\mathbb{N}$ contains nothing else but the familiar whole numbers.

8.4   Principle of induction

Suppose that a row of dominoes is arranged so that, if the *n*th one falls, it will push over the $(n+1)$th. Most people would agree that, if the first domino is now pushed over, then all the dominoes will fall down.

In arriving at this conclusion we have made an informal use of the principle of induction. The principle of induction is an immediate consequence of the definition of $\mathbb{N}$. Indeed, it is for this reason that the given definition is used. We begin with the following theorem.

8.5   *Theorem* Suppose that T is a set of natural numbers such that

(a) $1 \in T$
(b) $n \in T \Rightarrow n+1 \in T$.

Then $T = \mathbb{N}$.

Proof We have that $T \in \mathcal{W}$ and hence $\mathbb{N} \subset T$. But we are also given that $T \subset \mathbb{N}$. It follows that $T = \mathbb{N}$.

8.6   *Corollary* (*Principle of induction*) Suppose that n ranges over $\mathbb{N}$ and that $P(n)$ is a predicate. If

(a) $P(1)$,
(b) for any n, $P(n)$ implies $P(n+1)$,

are true statements, then $P(n)$ is true for any $n \in \mathbb{N}$.

Proof Simply take $T = \{n : P(n)\}$ in the preceding theorem.

8.7   Inductive definitions

An important use of the principle of induction is in making definitions. For example, when we write

$$x^n = x \cdot x \cdot \ldots \cdot x$$
$$\leftarrow n \text{ times} \rightarrow$$

what is really meant is that powers are to be defined inductively by the expressions

(a) $x^1 = x$
(b) $x^{n+1} = x^n \cdot x$ $(n \in \mathbb{N})$.

The principle of induction ensures that these expressions define x^n for all $n \in \mathbb{N}$. (See exercise 8.9(5).)

By this means, the notation '...' may be rendered respectable. It simply means that an inductive definition is to be used, the details of which are obvious.

Thus the notation

$$S_n = \sum_{k=1}^{n} a_k = a_1 + a_2 + \ldots + a_n$$

means that S_n is to be defined inductively by

(a) $S_1 = a_1$
(b) $S_{n+1} = S_n + a_{n+1}$ $(n \in \mathbb{N})$.

8.8   *Example* Prove that $1 + 2 + 3 + \ldots + n = \frac{1}{2} n(n+1)$.

Proof Put $S_n = 1 + 2 + 3 + \ldots + n$. This means that $S_1 = 1$ and $S_{n+1} = S_n + (n+1)$ for each $n \in \mathbb{N}$.

Let $P(n)$ be the predicate $S_n = \frac{1}{2} n(n+1)$. We shall prove that $P(n)$ is true for all $n \in \mathbb{N}$ by induction.

(a) Consider first $P(1)$. This asserts that $S_1 = \frac{1}{2} 1(1+1)$. Both sides are equal to 1 and hence $P(1)$ is true.

(b) Next we need to show that $P(n)$ implies $P(n+1)$ for any $n \in \mathbb{N}$. Choose any particular value of $n \in \mathbb{N}$ and assume that $P(n)$ is true – i.e. $S_n = \frac{1}{2} n(n+1)$. We need to deduce $P(n+1)$. But

$$\begin{aligned}
S_{n+1} &= S_n + (n+1) \\
&= \tfrac{1}{2} n(n+1) + (n+1) \\
&= \tfrac{1}{2} (n+1)(n+2)
\end{aligned}$$

and this is $P(n+1)$. Hence $P(n)$ implies $P(n+1)$.

We have therefore established items (a) and (b) in the principle of induction (corollary 8.6) and so the result follows.

8.9   *Exercise*

(1) Prove that

(i) $\displaystyle\sum_{k=1}^{n} k^2 = \tfrac{1}{6} n(n+1)(2n+1)$

(ii) $\displaystyle\sum_{k=1}^{n} k^3 = \tfrac{1}{4} n^2(n+1)^2$.

(2) Suppose that x and y are positive and that $n \in \mathbb{N}$. Prove that $x < y$ if and only if $x^n < y^n$.

(3) Prove that, if $x \neq 1$,

$$\sum_{k=0}^{n} x^k = 1 + x + x^2 + \ldots + x^n = \frac{1 - x^{n+1}}{1 - x}.$$

Hence show that, for any x and ξ,

$$x^n - \xi^n = (x - \xi)(x^{n-1} + x^{n-2}\xi + x^{n-3}\xi^2 + \ldots + \xi^{n-1}).$$

If $0 < \xi < x \leqq X$, deduce that $0 < x^n - \xi^n \leqq nX^{n-1}(x - \xi)$.

(4) Give an inductive definition of $n! = 1 \cdot 2 \cdot 3 \cdot \ldots \cdot n$ valid for all $n \in \mathbb{N}$. Assuming that $0! = 1$, define

$$\binom{n}{r} = \frac{n!}{r!(n-r)!} \quad (r = 0, 1, 2, \ldots, n).$$

Prove that

$$\binom{n}{r} + \binom{n}{r-1} = \binom{n+1}{r} \quad (r = 1, 2, \ldots, n).$$

Use this result and the principle of induction to obtain the binomial theorem in the form

$$(x + y)^n = \sum_{k=0}^{n} \binom{n}{k} x^{n-k} y^k.$$

†(5) Let ξ be any real number and let $\phi: \mathbb{R} \to \mathbb{R}$. Show that there exists a unique function $f: \mathbb{N} \to \mathbb{R}$ such that

    (a) $f(1) = \xi$
    (b) $f(n+1) = \phi(f(n))$ $(n \in \mathbb{N})$.

[*Hint*: Let $D_N = \{1, 2, 3, \ldots, N\}$. Use the principle of induction to show that, for each $N \in \mathbb{N}$, there exists a unique function $f_N: D_N \to \mathbb{R}$ satisfying (a) and (b) for $1 \leqq n < N$. Explain why f_{N+1} is an extension of f_N (exercise 6.4(6)). Now define $f: \mathbb{N} \to \mathbb{R}$ by $f(n) = f_n(n)$. Why is f the unique function satisfying (a) and (b)?]

†(6) Discuss the following argument which purports to show that all billiard balls are red. We begin with the induction hypothesis that all sets of n or less billiard balls contain only red balls. Now suppose we are given a set of $n+1$ billiard balls. This may be split into two non-empty subsets of n or less billiard balls. By the induction hypothesis both subsets contain only red balls. Putting the two subsets back together again, we conclude that any set of $n+1$ billiard balls or less contains only red balls. Hence the result by induction.

Construct similar fallacious arguments for the propositions

    (i) All billiard balls are the same colour.
    (ii) All points in the plane are collinear.

8.10 **Properties of** $\mathbb{N}$

We begin with a theorem which asserts that the sum and product of any pair of natural numbers are again natural numbers.

8.11 *Theorem* If m and n are natural numbers, then so are $m+n$ and mn.

Proof Regard $m \in \mathbb{N}$ as fixed and let $P(n)$ be the predicate $m+n \in \mathbb{N}$. We prove $P(n)$ by induction.

By theorem 8.3, $m \in \mathbb{N}$ implies $m+1 \in \mathbb{N}$ and so $P(1)$ is true. Next assume that $P(n)$ holds – i.e. $m+n \in \mathbb{N}$. It follows that $(m+n)+1 \in \mathbb{N}$ and thus $P(n+1)$ is true. This shows that $P(n)$ implies $P(n+1)$ for each $n \in \mathbb{N}$.

Now let $Q(n)$ be the predicate $mn \in \mathbb{N}$. The statement $Q(1)$ is simply $m \in \mathbb{N}$ and hence is true. Next assume that $Q(n)$ holds – i.e. $mn \in \mathbb{N}$. But $m(n+1) = mn + m \in \mathbb{N}$ by the previous paragraph. Thus $Q(n+1)$ is true. This shows that $Q(n)$ implies $Q(n+1)$ for each $n \in \mathbb{N}$.

From the axioms of the previous chapter we obtain the following properties of $\mathbb{N}$ which hold for all natural numbers l, m and n.

(I) *Arithmetic*

(i) $(l+m)+n = l+(m+n)$	(iv) $(lm)n = l(mn)$	(associative laws).
(ii) $l+m = m+l$	(v) $lm = ml$	(commutative laws).
(iii) $l+m = l+n \Leftrightarrow m = n$	(vi) $lm = ln \Leftrightarrow m = n$	(cancellation laws).
	(vii) $l(m+n) = lm + ln$	(distribution law).

(II) *Order*

(i) One and only one of the alternatives

$$m > n, \ m = n, \ m < n$$

holds.

(ii) $l+m > l+n \Leftrightarrow m > n$. (iii) $lm > ln \Leftrightarrow m > n$.

8.12 *Exercise*

(1) Let n be a natural number other than 1. Prove that $n-1$ is a natural number.

[*Hint*: Apply induction to the predicate '$n=1$ or $n-1 \in \mathbb{N}$'.]

If $m \in \mathbb{N}$, $n \in \mathbb{N}$ and $m > n$, prove that $m-n \in \mathbb{N}$.

[*Hint*: Apply induction to the predicate '$\forall m(m \leq n$ or $m - n \in \mathbb{N})$'.]

(2) Prove that every natural number $n \neq 1$ satisfies $n > 1$. If $n \in \mathbb{N}$, prove that there is no $x \in \mathbb{N}$ such that $n < x < n + 1$.

(3) A natural number n is even if $n = 2k$ for some natural number k; it is odd if $n = 1$ or $n = 2k + 1$ for some natural number k.

Prove that every natural number is either even or odd but never both.

(4) If m and n are natural numbers, prove that

(i) $x^{m+n} = x^m x^n$　(ii) $x^{mn} = (x^m)^n$　(iii) $(xy)^n = x^n y^n$

for all x and y. Show that m^n is a natural number. Prove that $2^n > n$ for all $n \in \mathbb{N}$.

†(5) If N is a natural number and $P(n)$ is a predicate such that the statements

(a) $P(N)$
(b) for any $n \in \mathbb{N}$ satisfying $1 \leq n < N$, $P(n+1) \Rightarrow P(n)$

are true, prove that $P(n)$ is true for each $n \in \mathbb{N}$ satisfying $1 \leq n \leq N$.

†(6) Let $P(n)$ be a predicate for which the statements

(a) $P(2)$
(b) for any $n \in \mathbb{N}$, $P(2^n) \Rightarrow P(2^{n+1})$
(c) For any $n \in \mathbb{N}$, $P(n+1) \Rightarrow P(n)$

are true. Prove that $P(n)$ is true for all $n \in \mathbb{N}$.

8.13　Integers

The set $\mathbb{Z}$ of integers is defined by

$$\mathbb{Z} = \{m - n : m \in \mathbb{N} \text{ and } n \in \mathbb{N}\}.$$

The properties of $\mathbb{Z}$ will be examined in chapter 11. At this stage we simply observe that $0, 1, -1, 2, -2, 3, -3$ are all examples of integers.

8.14　Rational numbers

The set $\mathbb{Q}$ of rational numbers (or fractions) is defined by

$$\mathbb{Q} = \left\{\frac{m}{n} : m \in \mathbb{Z} \text{ and } n \in \mathbb{N}\right\}.$$

If one adds, substracts, multiplies or divides two rational numbers, another rational number is obtained (provided that division by zero is excluded). For example,

$$\frac{2}{3} - \frac{7}{4} = \frac{8 - 21}{12} = -\frac{13}{12}$$

$$\frac{2}{3} \times \frac{7}{4} = \frac{14}{12} = \frac{7}{6}.$$

It follows that ℚ is an ordered field – i.e. the axioms of the previous chapter all remain valid with ℝ replaced by ℚ. Given that ℚ has all these nice properties, why then do we bother with ℝ?

To answer this question, it is necessary to recall that our motivation for studying the real numbers was to provide an idealisation of the notion of length in Euclidean geometry. At first sight, it certainly does seem that ℚ should be adequate for this purpose. One may imagine the process of marking all the rational numbers along a line. The integers would be marked first. Next would come the multiples of $\frac{1}{2}$: then the multiples of $\frac{1}{3}$ and so on. Assuming that this process could be completed, it seems difficult to see where there would be room for any further points on the line.

But this geometric intuition is not consistent with Pythagoras' theorem. This implies that the length h of the hypotenuse in the triangle below should satisfy

$$h^2 = 1^2 + 1^2 = 2.$$

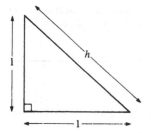

If it is true that every length in Eucilidean geometry can be measured by a rational number, then it must be true that there is a positive rational number such that $r = h$ and hence $r^2 = 2$. The next theorem shows that this is impossible.

8.15   *Theorem* There are no natural numbers m and n for which

$$m^2 = 2n^2.$$

Proof We suppose that m and n are natural numbers for which

$m^2 = 2n^2$ and seek a contradiction. From exercise 8.12(3), we know that m is either even or odd but not both.

Suppose that m is odd. Since $m \neq 1$, we have that $m = 2k + 1$ for some $k \in \mathbb{N}$. Hence $m^2 = (2k+1)^2 = 4k^2 + 4k + 1$ and thus m^2 is odd. But $m^2 = 2n^2$ and is therefore even. From this contradiction we conclude that m is even.

Since m is even, we can write $m = 2m_1$ for some $m_1 \in \mathbb{N}$. But then $4m_1^2 = m^2 = 2n^2$. Thus $2m_1^2 = n^2$. A repetition of the above argument yields that n is even and hence can be written in the form $n = 2n_1$ for some $n_1 \in \mathbb{N}$. Then $2m_1^2 = n^2 = 4n_1^2$ and so $m_1^2 = 2n_1^2$. A further repetition then shows that m_1 is even and so on.

It follows that both m and n can be divided by 2 as often as we choose. This is a contradiction. (Why?)

Theorem 8.15 tells us that, when all the rational numbers have been marked on a straight line, there will still be points on the line left unmarked. One can think of these points as 'holes' in the rational number system. In particular, there is a 'hole' where $\sqrt{2}$ ought to be.

The existence of these 'holes' makes the rational number system inadequate for measuring lengths. The real number system does not suffer from this deficiency. It contains extra objects (the irrational numbers) which fill up the 'holes' left by the rational numbers. When the real numbers are marked on the line, they fill it completely and stretch unbrokenly in both directions. We capture this idea by saying that the real number system forms a *continuum*.

To ensure the existence of these new 'irrational' numbers, we need a further axiom which we call the continuum axiom. This is the subject of the next chapter.

---

9 REAL NUMBERS (II)

9.1 Introduction

We began listing the axioms for the real number system in chapter 7. The axioms given were those of arithmetic and order. A system satisfying these axioms is called an ordered field. Thus the real number system is an ordered field. But this is not a full description of the real numbers. As we saw at the end of the last chapter, the system of rational numbers is another example of an ordered field.

To complete the description of the real number system, another axiom is required. We call this axiom the *continuum axiom*. Its purpose is to ensure that sufficient real numbers exist to fill up the 'holes' in the system of rational numbers (§8.14).

9.2 The method of exhaustion

This has nothing to do with getting your way at committee meetings. It is a method invented by Archimedes for finding the areas of regions with curved boundaries. We shall use the method to motivate the continuum axiom.

Briefly, the method involves packing polygons inside the region until its area is exhausted. In principle, one can calculate the area of any union of a finite number of the polygons. The area of the region in question is then identified with the smallest real number which is larger than each such polygonal area.

One of the shapes considered by Archimedes was the region trapped between a parabola and one of its chords. This example will be used to illustrate the idea. It is important to bear in mind that the argument to be used does not have the same status as those which we use when we prove theorems. The discussion is only intended to indicate why the theory might have interesting applications. We are therefore free to use geometrical ideas without constraint.

The area to be calculated is A. We begin by calculating the area of the triangle of A_1.

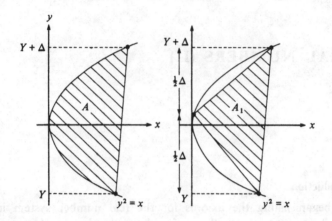

The calculation of A_1 requires a little co-ordinate geometry which we shall take for granted. The result obtained is

$$A_1 = (\tfrac{1}{4}\Delta)^3.$$

Notice that this area does not depend on the value of Y. We can therefore use the formula to help compute the areas of the polygons A_2, A_3 in the diagram below.

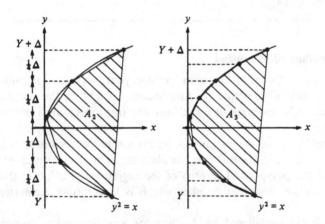

We obtain that

$$A_2 = (\tfrac{1}{4}\Delta)^3 + 2(\tfrac{1}{4}\cdot\tfrac{1}{2}\Delta)^3$$
$$A_3 = (\tfrac{1}{4}\Delta)^3 + 2(\tfrac{1}{4}\cdot\tfrac{1}{2}\Delta)^3 + 4(\tfrac{1}{4}\cdot\tfrac{1}{4}\Delta)^3.$$

In general,

$$A_n = (\tfrac{1}{4}\Delta)^3\left\{1 + (\tfrac{1}{2})^2 + (\tfrac{1}{4})^2 + \ldots + \left(\frac{1}{2^{n-1}}\right)^2\right\}$$

$$= (\tfrac{1}{4}\Delta)^3 \left\{ 1 + \tfrac{1}{4} + (\tfrac{1}{4})^2 + (\tfrac{1}{4})^3 + \ldots + (\tfrac{1}{4})^{n-1} \right\}$$

$$= (\tfrac{1}{4}\Delta)^3 \frac{1 - (\tfrac{1}{4})^n}{1 - \tfrac{1}{4}} = \frac{\Delta^3}{48} (1 - (\tfrac{1}{4})^n).$$

The smallest real number greater than A_n for all $n \in \mathbb{N}$ is $\Delta^3/48$ and hence this is the area of the region with which we are concerned.

Archimedes also considered the area of a circle and, using his method, was able to obtain as close an approximation to π as he chose. Consider the two sequences of polygons illustrated below.

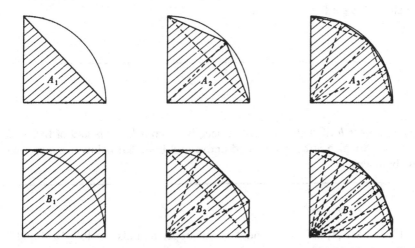

If the radius of the quarter-circle illustrated is 1, then $\tfrac{1}{4}\pi$ will be the smallest number larger than A_n for all $n \in \mathbb{N}$ and the largest number smaller than every B_n for all $n \in \mathbb{N}$. In particular, $A_n \le \tfrac{1}{4}\pi \le B_n$ for all n. But how does one calculate A_n and B_n for large values of n? Archimedes showed that $A_1 = \tfrac{1}{2}$, $B_1 = 1$ and

$$A_{n+1} = (A_n B_n)^{1/2} \qquad (n = 1, 2, \ldots)$$
$$B_{n+1}^{-1} = 2(A_{n+1}^{-1} + B_n^{-1}) \qquad (n = 1, 2, \ldots).$$

These recursive formulae make the calculations practicable.

The point of considering this example and the previous one is that they take for granted the existence of a smallest real number larger than each element of the set $\{A_n : n \in \mathbb{N}\}$. What else, after all, could the area of the curved region be? The continuum axiom is specifically designed to validate implicit assumptions of this sort.

9.3   Bounds

A set S of real numbers is *bounded above* if there exists a real number H such that $x \leq H$ for each $x \in S$.

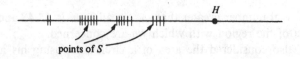

The number H (if such a number exists) is called an *upper bound* of the set S.

A set S of real numbers is *bounded below* if there exists a real number h such that $x \geq h$ for each $x \in S$.

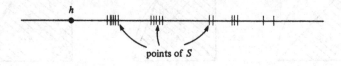

The number h (if such a number exists) is called a *lower bound* of the set S.

A set S which is both bounded above and bounded below is simply said to be *bounded*.

9.4   *Examples*

(i) The set $\{1, 2, 3\}$ is bounded. Some upper bounds are 100, 10, 4 and 3. Some lower bounds are -27, 0 and 1.

(ii) The interval $[1, 2) = \{x : 1 \leq x < 2\}$ is bounded. Some upper bounds are 100, 10, 4 and 2. Some lower bounds are -27, 0 and 1.

(iii) The interval $(0, \infty) = \{x : x > 0\}$ is *unbounded above*. If $H > 0$ is proposed as an upper bound, one has only to point to $H + 1$ as an element of the set larger than the supposed upper bound. However $(0, \infty)$ is bounded below. Some lower bounds are -27 and 0.

9.5   Continuum axiom

The continuum axiom (axiom V) is the last of our axioms for the real number system. An ordered field which satisfies this axiom is usually said to be *complete* but we shall prefer to describe an ordered field which satisfies axiom V as a *continuum ordered field*. In particular, the real number system is a continuum ordered field.

> ### Axiom V
> (i) Every non-empty set of real numbers which is bounded above has a *smallest* upper bound.
> (ii) Every non-empty set of real numbers which is bounded below has a *largest* lower bound.

If S is a non-empty set which is bounded above, the axiom asserts that S has an upper bound B such that, for any other upper bound H,

$$B \leq H.$$

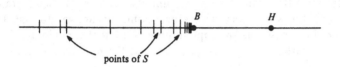

points of S

If S is a non-empty set which is bounded below, the axiom asserts that S has a lower bound b such that, for any other lower bound h,

$$b \geq h.$$

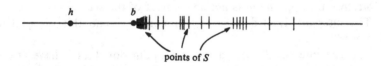

points of S

9.6   Examples

(i) The smallest upper bound of the set $\{1, 2, 3\}$ is 3. The largest lower bound is 1.

(ii) The smallest upper bound of the interval $[1, 2)$ is 2. The largest lower bound is 1.

(iii) The interval $(0, \infty)$ has no upper bounds at all. The largest lower bound is 0.

9.7   Supremum and infimum

A non-empty set S which is bounded above has a smallest upper bound, B, by the continuum axiom. This smallest upper bound is also called the *supremum* of the set S. We write $B = \sup S$ or

$$B = \sup_{x \in S} x.$$

Similarly, a non-empty set S which is bounded below has a largest lower bound, b. This is also called the *infimum* of the set S. We write $b = \inf S$ or

$$b = \inf_{x \in S} x.$$

It is of the greatest importance not to confuse these quantities with the maximum and the minimum of a set. The *maximum* of a set S is its largest element (if it has one). Similarly the *minimum* of a set S is its smallest element (if it has one). We use the notation max S and min S.

It is certainly true that, *if* the maximum of a set S exists, then it equals the supremum. But some sets which are bounded above do *not* have a maximum. For these sets the supremum exists but the maximum does not.

Similar remarks apply to the infimum and the minimum of a set.

9.8   *Examples*

(i) The set $\{1, 2, 3\}$ has a maximum 3 and this is equal to its supremum. The minimum of the set is 1 and this is equal to the infimum.

(ii) The interval $[1, 2) = \{x : 1 \leq x < 2\}$ has *no* maximum. Its supremum is 2. Note that the supremum is *not* an element of the set.

The interval $[1, 2)$ does have a minimum 1 and this is equal to the infimum.

(iii) The interval $(0, \infty)$ has no maximum nor does it have any upper bounds. Moreover, $(0, \infty)$ has *no* minimum. Its infimum is 0 which is *not* in the set $(0, \infty)$.

9.9   *Example* In theorem 8.15 it was shown that there exists no *rational* number q such that $q^2 = 2$. In this example we shall show that, for each $a > 0$, there exists a real number $\xi > 0$ such that $\xi^2 = a$. (In §9.13, a more general theorem is proved.)

Proof Let $L = \{l : l > 0 \text{ and } l^2 < a\}$ and $R = \{r : r > 0 \text{ and } r^2 > a\}$. We begin by showing that, for each $r \in R$, there exists an $s \in R$ such that $s < r$. Suppose that $r^2 > a$ and $r > 0$. Then $r > ar^{-1}$ and so $s = \frac{1}{2}(r + ar^{-1}) < r$. Also, $s^2 - a = (r^2 - a)^2/(4r^2) > 0$ and so $s \in R$. It follows that R has no minimum. A similar argument (involving $2(l^{-1} + la^{-1})^{-1}$) shows that L has no maximum.

Now let $\xi = \inf R$. Since R has no minimum, $\xi \notin R$. Thus $\xi^2 = a$ or else

$\xi \in L$. But each $l \in L$ is a lower bound for R and since ξ is the *largest* lower bound for R, $\xi \in L$ implies that $\xi = \max L$. But L has no maximum. It follows that $\xi^2 = a$ as required.

9.10 Exercise

(1) Suppose $x \leq K$ for each x in the non-empty set S. Prove that $\sup S \leq K$.

(2) *Prove* that the interval $(0, \infty)$ is unbounded above. *Prove* that $(0, \infty)$ has no minimum. *Prove* that its largest lower bound is 0.

(3) Let S be any non-empty set of real numbers and let T denote the set of all its upper bounds. Prove that either $T = \emptyset$ or else $T = [B, \infty)$ for some B.

(4) Let S be a non-empty set of real numbers which is bounded above. Prove that S has a maximum if and only if the supremum of S is an element of S.

(5) Let S be a non-empty set of real numbers which is bounded above and let $T = \{-x : x \in S\}$. Prove that H is an upper bound for S if and only if $-H$ is a lower bound for T. Deduce that T is bounded below and that $\inf T = -\sup S$. Explain why this result makes item (ii) of the continuum axiom redundant.

(6) *Prove* that every non-empty set of natural numbers has a minimum. [*Hint*: Use exercise 8.12(2).]

†(7) Let S be a non-empty set and let $f: S \to \mathbb{R}$ and $g: S \to \mathbb{R}$ have the property that $f(S)$ and $g(S)$ are bounded above. We use the notation

$$\sup_{x \in S} f(x) = \sup f(S).$$

Prove that

$$\sup_{x \in S} (f(x) + g(x)) \leq \sup_{x \in S} f(x) + \sup_{x \in S} g(x).$$

Give an example to show that $\leq$ cannot in general be replaced by $=$. Give a further example to show that the result need not be true if $+$ is replaced by $\times$. What if f and g do not take negative values?

†(8) Suppose that $b > 0$ and that S is bounded above. Prove that

$$\sup_{x \in S} (a + bx) = a + b \sup_{x \in S} x.$$

If $b < 0$, prove that

$$\sup_{x \in S} (a + bx) = a + b \inf_{x \in S} x.$$

†(9) Let S and T be two non-empty sets of real numbers which are bounded above. Prove that

$$\sup_{(x,y) \in S \times T} (x + y) = \sup_{x \in S} x + \sup_{y \in T} y.$$

If S and T are sets of positive numbers, show that the same result holds with $+$ replaced by $\times$.

†(10) Let S be a non-empty set of positive real numbers which is bounded above. If $n \in \mathbb{N}$, prove that

$$\sup_{x \in S} x^n = \left(\sup_{x \in S} x\right)^n.$$

†(11) Let S and T be non-empty sets and let $f : S \times T \to \mathbb{R}$ have bounded range. Prove that

$$\sup_{x \in S} \inf_{y \in T} f(x, y) \le \inf_{y \in T} \sup_{x \in S} f(x, y).$$

Show that $\le$ cannot in general be replaced by $=$. Show, however, that if $S = [0, 1]$, $T = [0, 1]$ and $f(x, y) = axy + bx + cy + d$, then $\le$ *can* be replaced by $=$. (This latter result is a special case of Von Neumann's 'maximum theorem' on which zero-sum game theory is based.)

†(12) Suppose that S and T are non-empty sets of real numbers which are bounded above. Prove that

(i) $S \subset T \Rightarrow \sup S \le \sup T$

(ii) $\sup (S \cup T) = \max\{\sup S, \sup T\}$.

What can be said about $\sup (S \cap T)$ and $\sup (\mathcal{C} S)$?

9.11†   Dedekind sections

We propose to use the real numbers for measuring distance in Euclidean geometry. For this purpose there must be a real number corresponding to each point on a straight line. What reason do we have to suppose that this is true?

Suppose that there is a point P which does *not* correspond to any real number. Geometric intuition tells us that P will split $\mathbb{R}$ into two disjoint subsets consisting of those real numbers to the left of P and those real numbers to the right of P. The left-hand set will have no maximum and the right-hand set will have no minimum.

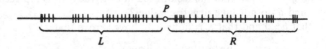

The theorem below shows that $\mathbb{R}$ cannot be split into two sets with these properties.

Consider first an ordering $\preccurlyeq$ on a set A (§5.8). A Dedekind section of A with respect to $\preccurlyeq$ is a pair $\{L, R\}$ of disjoint, non-empty subsets of A with the property that $A = L \cup R$ and

$$l \preccurlyeq r$$

for each $l \in L$ and $r \in R$. There are four possibilities to be distinguished which are classified in the diagram below.

	R has a minimum	R has no minimum
L has a maximum	jump	cut
L has no maximum	cut	gap

All Dedekind sections of $\mathbb{N}$ with respect to $\leq$ determine *jumps*. (See exercise 9.10(6).) As an example, consider the section $\{L, R\}$ of $\mathbb{N}$ in which $L = \{1, 2, 3\}$ and $R = \{4, 5, 6, \ldots\}$.

No Dedekind section of $\mathbb{Q}$ with respect to $\leq$ determines a jump. An example of a Dedekind section of $\mathbb{Q}$ which determines a *cut* is obtained by taking $L = (-\infty, \frac{1}{2}) \cap \mathbb{Q}$ and $R = [\frac{1}{2}, \infty) \cap \mathbb{Q}$. An example of a Dedekind section of $\mathbb{Q}$ which determines a *gap* is obtained by taking $L = \{r : r \in \mathbb{Q} \text{ and } r^3 \leq 2\}$ and $R = \{r : r \in \mathbb{Q} \text{ and } r^3 \geq 2\}$. The gap between these two sets has to be filled by the *irrational* number $2^{1/3}$.

If x is any real number, a Dedekind section $\{L, R\}$ of $\mathbb{R}$ with respect to $\leq$ can be obtained by taking $L = (-\infty, x)$ and $R = [x, \infty)$. Alternatively, one can take $L = (-\infty, x]$ and $R = (x, -\infty)$. The next theorem shows that *all* Dedekind sections of $\mathbb{R}$ are of one of these two simple types.

9.12† *Theorem* All Dedekind sections of $\mathbb{R}$ determine *cuts*.

Proof Let $\{L, R\}$ be a Dedekind section of $\mathbb{R}$. If L has no maximum, then R is the set of all upper bounds of L and hence has a minimum by the continuum axiom. This completes the proof.

9.13† **Powers**

In §8.7, we considered powers x^n for which n is a natural number. We now propose to consider how to proceed when n is *not* a natural number.

9.14† *Theorem* Suppose that $y > 0$. Then there exists a *positive* real number x such that

$$x^n = y.$$

Proof Define $L = \{x : x > 0 \text{ and } x^n < y\}$ and $R = \{x : x > 0 \text{ and } x^n > y\}$. If the theorem is false, then $L \cup R = (0, \infty)$. Thus $\{L, R\}$ is a Dedekind section of $(0, \infty)$ and hence determines a cut (exercise 9.15(1)). This is a contradiction since L has no maximum and R has no minimum.

We prove that L has no maximum. Let $\xi \in L$ and $X \in R$. Then $\xi^n < y$ and $X^n > y$. If

$\xi < x \leq X$, it follows from exercise 8.9(3) that

$$0 < x^n - \xi^n \leq nX^{n-1}(x - \xi).$$

Thus $x^n < y$ and so $x \in L$, provided that $x - \xi$ is sufficiently small. (See exercise 9.15(8).) Thus, for each $\xi \in L$, there is a larger $x \in L$. A similar argument shows that R has no minimum.

If $y > 0$, the positive real number $x > 0$ which satisfies $x^n = y$ is *unique* (exercise 8.9(2)). We use the notation

$$x = y^{1/n}$$

and call x the nth *root* of y. Note that, with this convention, the nth root of a positive y is *always* positive. In particular, the notation $\sqrt{y} = y^{1/2}$ stands for the *positive* x such that $x^2 = y$. There are, of course, two numbers whose square is y. The positive one is $\sqrt{y}$ and the negative one is $-\sqrt{y}$. The notation $x = \pm\sqrt{y}$ means '$x = \sqrt{y}$ or $x = -\sqrt{y}$.'

If $r = m/n$ is a positive rational number and $y > 0$, we define

$$y^r = (y^m)^{1/n}.$$

If r is a negative rational number, we define

$$y^r = \frac{1}{y^{-r}}.$$

We also define $y^0 = 1$ and $0^r = 0$ provided that $r > 0$. (A general discussion of the meaning of y^r when $y < 0$ requires the use of complex numbers.)

9.15†   *Exercise*

(1) Prove that any Dedekind section of $(0, \infty)$ determines a cut.
(2) Prove that, for any rational numbers r and s, there is a rational number t satisfying $r < t < s$. Deduce that no Dedekind section of $\mathbb{Q}$ determines a jump.
(3) Find all Dedekind sections of the set

$$A = (0, 1) \cup (1, 2) \cup [2, 3] \cup (4, 5] \cup \{6\}$$

with respect to $\leq$ which are not cuts.
(4) Suppose that n is an *even* natural number. Prove that the equation $x^n = y$ has no (real) solutions if $y < 0$, one solution if $y = 0$ and two solutions if $y > 0$. If n is an *odd* natural number, prove that the equation always has exactly one (real) solution.
    Sketch the graphs of $y = x^2$ and $y = x^3$ to illustrate these results.
(5) Simplify the following expressions:

    (i) $8^{2/3}$ (ii) $27^{-4/3}$ (iii) $32^{6/5}$.

(6) If $y > 0$, $z > 0$ and r and s are any rational numbers, prove that

    (i) $y^{r+s} = y^r y^s$ (ii) $y^{rs} = (y^r)^s$ (iii) $(yz)^r = y^r z^r$.

(7) Suppose that $a>0$ and $b^2-4ac>0$. Prove that

$$ax^2+bx+c=a(x-\alpha)(x-\beta)$$

for all $x\in\mathbb{R}$, where

$$\alpha=\frac{-b+\sqrt{b^2-4ac}}{2a}, \quad \beta=\frac{-b-\sqrt{b^2-4ac}}{2a}.$$

Deduce that $ax^2-bx+c<0$ if and only if $\alpha<x<\beta$. Find the minimum value of ax^2+bx+c without the use of calculus.

(8) In the proof of theorem 9.14 show that x must be chosen such that

$$\xi<x<\xi+\frac{1}{nX^{n-1}}(y-\xi^n).$$

Confirm that the right-hand side is less than X.

(9) Prove the inequality of the arithmetic and geometric means – i.e. if $a_1, a_2, \ldots, a_n$ are positive, prove that

$$\{a_1a_2\ldots a_n\}^{1/n}\leq\frac{1}{n}(a_1+a_2+\ldots+a_n).$$

[*Hint*: Use exercise 8.12(6).]

9.16 Infinity

Are ∞ and $-\infty$ real numbers? This is obviously not the case if one insists that these are the *only* 'infinite' elements in $\mathbb{R}$. It would then (presumably) be necessary to require that $\infty+\infty=\infty$. But, if it were true that ∞ is a real number, it would have to satisfy the usual rules of arithmetic. In particular, from $\infty+\infty=\infty$ we could deduce the nonsensical result

$$\infty=\infty-\infty=0.$$

A much more plausible heresy is to postulate the existence of *many* 'infinite' elements. Indeed, Isaac Newton's original development of the calculus was based on just this idea and required the existence of infinitesimals (i.e. 'infinitely small' but non-zero objects) which he called fluxions. But the existence of such objects is incompatible with the continuum axiom. There are *no* infinite or infinitesimal real numbers. If one wishes to entertain the existence of such objects it is necessary to work in some different number system for which the continuum axiom is false.

In order to discuss this matter in a rational way, we must first remove it from the realm of metaphysics by saying precisely what we mean by an 'infinite' element. Otherwise our discussion will resemble the medieval controversy about the number of angels who can simultaneously dance on the end of a pin.

We know from the previous chapter that the system ℝ of real numbers contains the subset ℕ of natural numbers. A reasonable definition of an infinite element of ℝ is an element which is larger than each natural number. Such an infinite element of ℝ (if it exists) is therefore an upper bound for ℕ.

The question of whether ℝ has infinite elements then reduces to the question of whether ℕ is bounded above.

9.17   *Theorem (Archimedean postulate)* The set ℕ is unbounded above.

Proof Suppose that ℕ is bounded above. By the continuum axiom, it then has a smallest upper bound B. Since B is the *smallest* upper bound of ℕ, $B-1$ is not an upper bound. Hence there exists $n \in \mathbb{N}$ such that $n > B - 1$ and therefore

$$n + 1 > B.$$

But $n \in \mathbb{N}$ implies $n + 1 \in \mathbb{N}$. We have therefore found an element of ℕ larger than the upper bound B. This is a contradiction.

The fact that ℕ is unbounded above is called the *Archimedean property* of the real number system ℝ. An ordered field for which the equivalent property is false, is therefore called *non-Archimedean*. In such ordered fields, infinite and infinitesimal elements *do* exist. These are studied in 'non-standard analysis'.

9.18   *Exercise*

(1) Prove that the largest lower bound of the set $S = \{1/n : n \in \mathbb{N}\}$ is 0. Explain what this has to do with the existence of infinitesimal elements of ℝ.
(2) Prove that the set $\{2^n : n \in \mathbb{N}\}$ is unbounded above.
(3) Prove that any non-empty set of integers which is bounded above has a maximum and any set of integers which is bounded below has a minimum. (See exercise 9.10(6).)

9.19   Denseness of the rationals

Let a and b be two real numbers with $a < b$. We can always find a real number x between a and b. The obvious example of such an x is

$$x = \tfrac{1}{2}(a + b).$$

This, incidentally, shows that it makes no sense to talk about two real numbers being 'next to each other'. If a and b are not equal, there is a real number between them and so they cannot be 'next to each other'.

It is not so obvious that there is a *rational* number between every pair of real numbers. But this is true. We express the fact by saying that $\mathbb{Q}$ is *dense* in $\mathbb{R}$.

9.20   *Theorem* Let a and b be real numbers with $a < b$. Then there exists a rational number r with $a < r < b$.

Proof Since $\mathbb{N}$ is unbounded above, we can find an $n \in \mathbb{N}$ such that

$$n > \frac{1}{(b-a)}$$

i.e.

$$\frac{1}{n} < b - a.$$

Let m be the smallest integer such that

$$r = \frac{m}{n} > a.$$

Then

$$\frac{m-1}{n} \leqq a.$$

Hence

$$a < \frac{m}{n} = \frac{m-1}{n} + \frac{1}{n} < a + (b-a) = b.$$

9.21† Uniqueness of the real numbers

The axioms we have given for the real number system are *categorical*. This means that they are satisfied by only one mathematical structure – i.e. the system $\mathbb{R}$ is *unique*.

It is quite important to understand the meaning of the word 'unique' in this context. It does not necessarily mean that the real numbers have unique *names*. For example, we now use the symbol 6 where the Romans used the symbol VI. But this does not mean that we are talking about a different number. The algebraists have a special word to describe this situation. They say that two systems are *isomorphic* if they have precisely the same mathematical structure and differ only in the symbols used to denote the various objects and relations operating in the system.

When we say that the real number system is unique, what is really meant is that all systems which satisfy the axioms for the real number system are isomorphic. Classicists using Roman numeration and computer engineers using binary notation are therefore to be regarded as using the same number system as ourselves. On the other hand, the Illinois State Legislature is said to have passed a law requiring that $\pi = 3$ (since this is what it says in the Bible). The mathematical structure of the Illinois number system is therefore very definitely not the same as ours.

How can we show that all systems which satisfy the axioms for the real number system are isomorphic? To discuss this point, we first need to be more specific about the word 'isomorphic'.

Suppose that $\mathbb{R}_1$ and $\mathbb{R}_2$ are two ordered fields. Then we say that $\mathbb{R}_1$ and $\mathbb{R}_2$ are *isomorphic* if and only if there exists a bijection $f: \mathbb{R}_1 \to \mathbb{R}_2$ such that

(i) $f(x+y) = f(x) + f(y)$
(ii) $f(xy) = f(x)f(y)$
(iii) $x \leq y \Leftrightarrow f(x) \leq f(y)$.

The function f is called an *isomorphism*. If $\mathbb{R}_1$ and $\mathbb{R}_2$ are isomorphic, it is sensible to say that they are really the same system except that the object labelled by x in $\mathbb{R}_1$ is labelled by $f(x)$ in $\mathbb{R}_2$.

Now suppose that $\mathbb{R}_1$ and $\mathbb{R}_2$ are ordered fields which satisfy the continuum axiom. We may construct, as in chapter 8, a subsystem $\mathbb{Q}_1$ of $\mathbb{R}_1$ and a subsystem $\mathbb{Q}_2$ of $\mathbb{R}_2$. We shall take for granted the more or less 'obvious' proposition that $\mathbb{Q}_1$ and $\mathbb{Q}_2$ are isomorphic. How does one use this fact to construct an isomorphism $f: \mathbb{R}_1 \to \mathbb{R}_2$? One way of proceeding is to use Dedekind sections.

Let $\mathcal{D}_1$ denote the set of all Dedekind sections $\{L, R\}$ of $\mathbb{Q}_1$ for which L has no maximum and consider the function $f_1: \mathbb{R}_1 \to \mathcal{D}_1$ defined by $\{L, R\} = f_1(x)$ where

$$L = \{r: r \in \mathbb{Q}_1 \text{ and } r < x\}$$
$$R = \{r: r \in \mathbb{Q}_1 \text{ and } r \geq x\}.$$

A simple consequence of theorem 9.19 (see exercise 9.22(3)) is that $f_1: \mathbb{R}_1 \to \mathcal{D}_1$ is a bijection. If we define $f_2: \mathbb{R}_2 \to \mathcal{D}_2$ in a similar way this will also, of course, be a bijection.

Next we introduce an isomorphism $\phi: \mathbb{Q}_1 \to \mathbb{Q}_2$. This can be used to define a bijection $\Phi: \mathcal{D}_1 \to \mathcal{D}_2$ by writing

$$\Phi(\{L, R\}) = \{\phi(L), \phi(R)\}.$$

The bijection $f: \mathbb{R}_1 \to \mathbb{R}_2$ defined by $f = f_2^{-1} \circ \Phi \circ f_1$ is then an isomorphism between $\mathbb{R}_1$ and $\mathbb{R}_2$.

We shall not provide a formal proof of the fact that f is an isomorphism since this would anticipate some of the work of the next chapter. The essential point is that the real numbers and their properties are entirely determined by the rational numbers and their properties. In particular, there is precisely one irrational number for each gap in the rationals and there are no other irrational numbers.

All the above analysis takes for granted, of course, that the axioms for the real

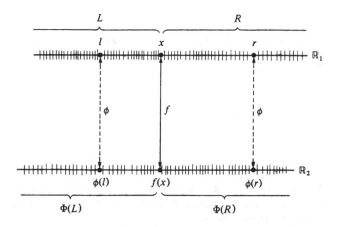

number system are *consistent* and so there is at least one mathematical structure which satisfies them. We take this question up in the next chapter.

9.22† *Exercise*

(1) Let x be a real number. Prove that x is the smallest upper bound of the set of all rational numbers less than x and the largest lower bound of all rational number greater than x.

(2) Let $\{L, R\}$ be a Dedekind section of $\mathbb{Q}$. Prove that $\sup L = \inf R$.

(3) Show that the function $f : \mathbb{R}_1 \to \mathbb{R}_2$ of §9.21 is a bijection.

(4) Suppose that $0 < a < b$. Prove that there exists a rational number r such that $a < r^2 < b$. [*Hint*: Show that, if n is sufficiently large, an m can be found such that $a < (m/n)^2 < b$. If $h^2 > a$ (e.g. $h = a + 1$), n will be large enough if $n > (2h + 1)(b - a)^{-1}$.] Let L be the set of all positive rational numbers r satisfying $r^2 \leq 2$ and let R be the set of all positive rational numbers satisfying $r^2 \geq 2$. Show that $\{L, R\}$ determines a gap in $\mathbb{Q}$ and deduce the existence of the unique irrational $\sqrt{2}$.

10† CONSTRUCTION OF THE NUMBER SYSTEMS

10.1† Models

A model for a list of axioms is an example of a system in which the axioms hold. It is the models in terms of which a list of axioms can be interpreted which make the axioms interesting.

But models can also be useful in examining the structure of systems of axioms. A famous example relates to Euclid's parallel postulate. Many mathematicians tried for a very long time to deduce this postulate from the other axioms of Euclidean geometry. All attempts were unsuccessful because the project is *impossible*. Models can be constructed in which all the other axioms of Euclidean geometry are true but the parallel postulate is *false*.

In this chapter, we construct a model for the real number axioms. The primary reason for producing this model is that its existence demonstrates that the axioms given for the real number system are *consistent* – i.e. they do not lead to a contradiction. It need hardly be said that an inconsistency in the axioms for the real number system would be a total disaster for the whole of analysis and most other branches of mathematics as well.

We have already seen (§9.21) that there is essentially only *one* model of the real number system – i.e. all models have precisely the same structure and differ only in the symbols which are used to label their elements. It therefore makes sense to identify a real number with the object we shall construct to represent it – i.e. to say that our model *is* the real number system. This is, in fact, the traditional attitude to the real numbers and the construction of our model mimics the historical process by which the real numbers were discovered. We can therefore combine our analysis of the consistency of the real number axioms with a discussion of how the real number system came to be invented as the culmination of a progression of number systems of ever-increasing power. To some extent this discussion reverses the treatment of chapter 8. Instead of constructing $\mathbb{N}$ from $\mathbb{R}$, we shall begin with $\mathbb{N}$ and construct $\mathbb{R}$. But the two approaches should not be regarded as competitors. Each approach has insights to offer that the other does not.

A fully-detailed account of the constructive approach would be very lengthy. We therefore confine the text to an account of the basic ideas together with the proofs of typical theorems. The remaining theorems have been listed as exercises. It is not suggested, however, that the reader should necessarily attempt *all* of these exercises. It would be more sensible to read through the exercises to see what they say and to attempt the proofs of a representative sample. A more detailed account of the theory is to be found in E. Landau's *Foundations of Analysis* (Chelsea, 1951).

10.2† Basic assumptions

Houses cannot be built without bricks and mortar. Nor can theorems be proved without axioms to base them on. We shall therefore get nowhere at all in constructing the real number system unless we have some basic assumptions with which to work.

In the construction, all the objects will be defined as sets and our basic assumptions will be the usual rules for forming and manipulating these sets (see chapter 4). In addition, we shall require one further assumption called the *axiom of infinity*. This is discussed in the next section.

10.3† Natural numbers

The first step is to construct the natural numbers. We first encounter these when, as children, we are taught to count. The essence of counting is that, for each natural number, there is a new and different natural number which follows it. This new natural number can then be used to label the next object in the set we are counting. Our problem is somehow to capture this primitive idea without importing any preconceived ideas about the natural numbers which we may have inherited from the classroom.

Everything in our construction is to be a set. Our initial aim is therefore to find some suitable sets to serve as natural numbers. It seems elegant to choose a set to represent n which we would intuitively regard as having n elements. In particular, we want a set containing just one element to serve as 1. It is usual to take

$$1 = \{\emptyset\}$$

– i.e. 1 is the set whose unique element is the empty set. We can then define

$$1 = \{\emptyset\}$$
$$2 = \{\emptyset, 1\}$$
$$3 = \{\emptyset, 1, 2\}$$
$$4 = \{\emptyset, 1, 2, 3\}$$

and so on. The rule for finding the natural number which follows n is simple to state. If n is a natural number, its *successor* $s(n)$ is simply

$$s(n) = n \cup \{n\}.$$

Thus $2 = 1 \cup \{1\} = \{\emptyset\} \cup \{1\} = \{\emptyset, 1\}$ and $3 = 2 \cup \{2\} = \{\emptyset, 1\} \cup \{2\} = \{\emptyset, 1, 2\}$.

The discussion above provides a precise formal definition for any particular natural number. It does *not* provide a precise formal definition of the set of *all* natural numbers. For this purpose we require the *axiom of infinity*. This asserts the existence of at least one set S satisfying

(i) $1 \in S$
(ii) $x \in S \Rightarrow x \cup \{x\} \in S$

for each $x \in S$. The collection $\mathcal{W}$ of all such sets S is therefore non-empty and so we can define the set $\mathbb{N}$ of natural numbers by

$$\mathbb{N} = \bigcap_{S \in \mathcal{U}} S.$$

10.4   Peano postulates

The definition given above for $\mathbb{N}$ is very similar to that of §9.2. The only difference is that instead of $x + 1$ we have $s(x) = x \cup \{x\}$. However, this difference does not affect the proof of the principle of induction (corollary 8.6) which we obtain in the following form. Suppose that $P(n)$ is a predicate for which the statements

(a) $P(1)$
(b) for any $n \in \mathbb{N}$, $P(n) \Rightarrow P(s(n))$

are true. Then $P(n)$ is true for any $n \in \mathbb{N}$.

Having made this observation, we can now state the Peano postulates for $\mathbb{N}$.

(i) $1 \in \mathbb{N}$.
(ii) There exists an injective function
$s: \mathbb{N} \rightarrow \mathbb{N} \setminus \{1\}$.
(iii) The principle of induction holds.

In what follows, we shall forget where $\mathbb{N}$ came from and simply use the Peano postulates. This makes two things clear. The first is that the axiom of infinity is just a disguised version of the principle of induction. The principle of induction is therefore a fundamental assumption for the constructive approach. The second point to note is that the definition given for $\mathbb{N}$ in §10.3 is essentially arbitrary. Almost any other definition would have done as well provided that it led to a suitable successor function $s: \mathbb{N} \rightarrow \mathbb{N} \setminus \{1\}$. We would then, of course, have introduced whatever different axiom of infinity was necessary in order for it to be possible to deduce the principle of induction.

10.5   *Example* Prove that each $n \in \mathbb{N}$ other than 1 has a predecessor – i.e. $s: \mathbb{N} \rightarrow \mathbb{N} \setminus \{1\}$ is surjective.

Proof Let $P(n)$ be the predicate '$n = 1$ or $n = s(m)$ for some $m \in \mathbb{N}$'. Clearly $P(1)$ is true. Suppose that $P(n)$ holds. We then have to deduce the truth of $P(s(n))$. But this is immediate since n is a predecessor for $s(n)$. The result therefore follows by induction.

10.6†   Arithmetic and order

We shall now use the Peano postulates to provide inductive definitions for addition and multiplication in $\mathbb{N}$.

For each $n \in \mathbb{N}$, we *define*

$$n + 1 = s(n).$$

Now this definition has been made, we can abandon the clumsy notation $s(n)$ and use $n+1$ instead. If we assume that $n+m$ is defined, we may then define $n+(m+1)$ by the formula

$$n + (m+1) = (n+m) + 1.$$

Since we are assuming the principle of induction, this defines $n+m$ for all $m \in \mathbb{N}$. In particular, we have that

$$\begin{aligned}
2 + 2 &= 2 + (1+1) \\
&= (2+1) + 1 \\
&= 3 + 1 \\
&= 4.
\end{aligned}$$

Similarly, if $n \in \mathbb{N}$, then we define $n1$ by

$$n1 = n$$

If we assume that nm has been defined, we may then define $n(m+1)$ by the formula

$$n(m+1) = nm + n.$$

It follows from the principle of induction that nm is then defined for all $m \in \mathbb{N}$. In particular.

$$\begin{aligned}
2 \times 2 &= 2(1+1) \\
&= 2 \times 1 + 2 \times 1 \\
&= 2 + 2 \\
&= 4.
\end{aligned}$$

Finally, we define the meaning of $m > n$. We say that $m > n$ if and only if $m = n + k$ for some $k \in \mathbb{N}$. In particular,

$$2 = 1 + 1 > 1.$$

The next step is to deduce all of the arithmetic and order properties given for $\mathbb{N}$ in §8.10 using only the Peano postulates and the inductive definitions for addition and multiplication introduced above.

In the following examples, we illustrate the techniques by means of which these and other properties for the natural numbers may be proved and leave the rest of the work as an exercise.

10.7   *Example* Prove that $(l+m) + n = l + (m+n)$.

Proof Let $P(n)$ be the predicate $(l+m) + n = l + (m+n)$. Then $P(1)$ is simply $(l+m) + 1 = l + (m+1)$ and this is true *by definition*. Now suppose that $P(n)$ is true and consider

$$(l+m)+(n+1) = \{(l+m)+n\}+1 \quad \text{(definition)}$$
$$= \{l+(m+n)\}+1 \quad \text{(since } P(n) \text{ is true)}$$
$$= l+\{(m+n)+1\} \quad \text{(definition)}$$
$$= l+\{m+(n+1)\} \quad \text{(definition)}.$$

Thus we have shown that $P(n) \Rightarrow P(n+1)$ and so the result follows by induction.

10.8   *Example* Let $l \in \mathbb{N}$. Show that there exists no $m \in \mathbb{N}$ such that $l = m+l$. (Thus $\mathbb{N}$ contains no 'zero'.)

Proof Because $1 \notin s(\mathbb{N})$ we have that $1 \neq m+1$. Suppose that

$$l+1 = m+(l+1)$$

i.e. $l+1 = (m+l)+1$ (definition).

Since s is injective it follows that

$$l = (m+l).$$

Hence $l \neq m+l \Rightarrow (l+1) \neq m+(l+1)$ and the result follows by induction.

10.9†   *Exercise*

(1) Prove that every natural number $n \neq 1$ satisfies $n > 1$.
(2) Establish the properties of $\mathbb{N}$ listed in §8.10. [*Hint*: A convenient order in which to prove the properties is Ii, Iii, Iv, Ivii, Iiv, IIi, IIii, IIiii, Iiii, Iiv. Note that Ii has already been proved (example 10.7). For properties Iii and Iv it may be helpful to prove first of all that $l+1 = 1+l$ and $l1 = 1l$ by induction on l. Finally observe that, once IIi has been proved, Iiii follows from IIii and Ivi from IIiii.]
(3) Let m and n be natural numbers. Show that the equation

$$m = n+x$$

admits a solution $x \in \mathbb{N}$ if and only if $m > n$. [*Hint*: See exercise 8.12(1).] If $m > n$, prove that the solution x is unique. We write $x = m-n$. Prove that

$$(l+m)-n = l+(m-n)$$

provided $m > n$.
(4) Let m and n be natural numbers. If the equation

$$m = nx$$

admits a solution $x \in \mathbb{N}$, we say that 'n divides m'. If n divides m, prove that the solution x is unique. We write $x = m/n$. Prove that

$$\left(\frac{l}{m}\right)n = \frac{(ln)}{m}$$

provided m divides l.

(5) If $n \in \mathbb{N}$, prove there is no $x \in \mathbb{N}$ such that $n < x < n+1$.

(6) If $m \in \mathbb{N}$ and $n \in \mathbb{N}$, then $m \leq n$ means '$m < n$ or $m = n$'. Show that $\leq$ is an ordering on $\mathbb{N}$ in the sense of §5.8. Prove that every non-empty set of natural numbers has a minimum with respect to this ordering.

10.10† Measuring lengths

The motivation for introducing the natural numbers is obvious. We need them to count quantities like herds of cows which come in discrete units. But some quantities such as length or time do not come in discrete units.

The natural way to proceed in measuring length is to begin with a measuring rod which will represent our unit of measurement. If duplicates of the measuring rod are available, they can be laid end to end alongside the length to be measured. If an exact fit is obtained with 4 rods, we say that the length is 4 units long.

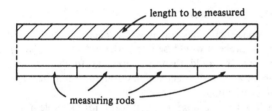

If 3 lengths each 4 units long are laid end to end, we shall need 12 measuring rods to obtain an exact fit to the total length.

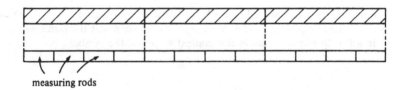

Observe that

$$12 = 3 \times 4.$$

This gives us a clue on how to proceed in the general case when it is impossible to obtain an exact fit with a whole number of measuring rods. If we find that m measuring rods are necessary to obtain an exact fit when n copies of the length to be measured are laid end to end, then we shall wish to say that the required length is x where

$$m = nx.$$

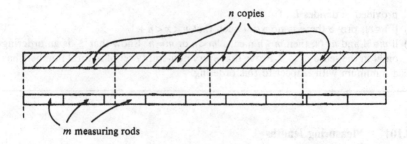

m measuring rods

Unfortunately, if we are restricted to working only with natural numbers, we have to say that the equation $m = nx$ usually has no solutions. The Greeks avoided this problem by simply asserting that the length and the measuring rod are in the ratio $m : n$. The modern approach is to *construct* a system of numbers in which every equation $m = nx$ has a solution. This system is the system $\mathbb{Q}_+$ of positive rational numbers. How is this system to be constructed from $\mathbb{N}$?

10.11† Positive rational numbers

We wish to construct a system in which each equation $m = nx$ has a solution. A natural way to proceed would be to use the ordered pair (m, n) to stand for the solution of $m = nx$. It would then be necessary to define addition and multiplication on the set $\mathbb{N} \times \mathbb{N}$ of all ordered pairs of natural numbers in a manner compatible with this interpretation.

But this approach will not do because we shall want, for example, the equations $m = nx$ and $2m = 2nx$ to have the *same* solution but (m, n) and $(2m, 2n)$ are not the same ordered pair. We therefore introduce an equivalence relation $\sim$ on $\mathbb{N} \times \mathbb{N}$ by writing

$$(m, n) \sim (p, q) \Leftrightarrow mq = pn.$$

This equivalence relation splits $\mathbb{N} \times \mathbb{N}$ into disjoint equivalence classes (theorem 5.7) and it is these equivalence classes we shall use to represent the elements of $\mathbb{Q}_+$.

It will help our intuition in this context if we use the symbol

$$\frac{m}{n}$$

to label the equivalence class which contains the order pair (m, n). We then need to note that

$$\frac{m}{n} = \frac{p}{q} \Leftrightarrow mq = pn.$$

Of course, the mere act of choosing the notation m/n does not advance the mathematics of the situation at all (and, unless we are careful, may even prove a hindrance by tempting us to use the properties of fractions before we have established them).

The natural numbers are fitted inside the system by making the identification

$$\frac{m}{1} = m.$$

Addition, multiplication and an ordering are defined by

$$\frac{m}{n} + \frac{p}{q} = \frac{mq = pn.}{nq}$$

$$\frac{m}{n} \times \frac{p}{q} = \frac{mp}{nq}$$

$$\frac{m}{n} > \frac{p}{q} \iff mq > np.$$

We can now check that m/n is the unique solution to the equation $m = nx$ when m and n are natural numbers. Consider the equation

$$\frac{m}{1} = \frac{n}{1} \times \frac{p}{q}.$$

By definition,

$$\frac{n}{1} \times \frac{p}{q} = \frac{np}{q}$$

and so

$$\frac{m}{1} = \frac{np}{q} \iff mq = np \iff \frac{m}{n} = \frac{p}{q}.$$

We have now constructed $\mathbb{Q}_+$ as the set of all objects of the form m/n with equality, addition, multiplication and an ordering defined as above. Next comes the task of deducing the properties of $\mathbb{Q}_+$ from the properties of $\mathbb{N}$ which we have already established and the given definitions. Since we have identified $\mathbb{N}$ with a subset of $\mathbb{Q}_+$, we must be particularly careful to check that our definitions for addition etc. in $\mathbb{Q}_+$ agree with those already established for $\mathbb{N}$. These tasks are the subject of the next set of exercises.

10.12† *Exercise*

(1) Prove that the relation $\sim$ introduced in §10.11 is an equivalence relation on $\mathbb{N} \times \mathbb{N}$ – i.e. it satisfies the rules of §5.5.

(2) Prove that the definitions of addition, multiplication and the ordering introduced in §10.11 are consistent with the criterion for equality – i.e. if $x_1 = x_2$ and $y_1 = y_2$, check that

$$x_1 + y_1 = x_2 + y_2$$

$$x_1 y_1 = x_2 y_2$$

$$x_1 < y_1 \iff x_2 < y_2.$$

(3) Check that the identification $m/1 = m$ is a sensible one in the following way.

Show that

$$m/n = k$$

where k is a natural number if and only if n divides m and, in that case, the definition of m/n agrees with that discussed in exercise 10.9(4). Check that the definitions for addition, multiplication and the ordering are consistent with those of $\mathbb{N}$ in the case when $n = q = 1$.

(4) Check that the arithmetical rules listed in §8.10 also hold in the case when l, m and n belong to $\mathbb{Q}_+$. Prove also that $\mathbb{Q}_+$ is a *group* under multiplication. If $s > t$, show that there is a unique $r \in \mathbb{Q}_+$ such that $s = t + r$. (We write $r = s - t$.)

(5) Check that the definition given for an ordering on $\mathbb{Q}_+$ satisfies the rules for an ordering listed in §5.8. (Here, of course, $x \leq y$ is to be interpreted as '$x < y$ or $x = y$'.) Check also that the ordering rules listed in §8.10 hold in the case when l, m and n are elements of $\mathbb{Q}_+$.

(6) Prove that, for each $r \in \mathbb{Q}_+$ and $s \in \mathbb{Q}_+$, there exists a *unique* $x \in \mathbb{Q}_+$ such that

$$r = sx.$$

10.13†   Positive real numbers

We can approximate to lengths as closely as we like using positive rational numbers. But if we wish to be able to measure all lengths *exactly*, we need to turn to the positive *real* numbers. This point was discussed thoroughly in chapters 8 and 9. How do we *construct* the system $\mathbb{R}_+$ of positive real numbers from $\mathbb{Q}_+$?

The idea is a simple one first introduced by the Greek Eudoxus but put into its modern powerful form by the German mathematician Dedekind. The essentials of the idea have already been discussed in §9.2. Given any 'point' on a line, consider the set of positive rational numbers which lie to its 'left'. We shall call this set a *section* and use it as a label or name for the 'point' with which we started. In this way we can discuss the 'points' on the line entirely in terms of the properties of $\mathbb{Q}_+$.

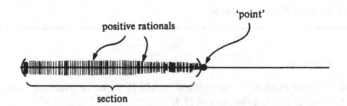

Our problem now is to convert the idea above into precise mathematical language.

We begin by defining a *section* of $\mathbb{Q}_+$ to be a non-empty set L of positive rational numbers which satisfies:

   (i) if $s \in L$ and $r < s$, then $r \in L$
  (ii) L is bounded above
 (iii) L has no maximum.

Item (ii) means, of course, that for some positive rational number t, $s \le t$ for each $s \in L$. Item (iii) says that no such t actually belongs to the set L.

We now identify $\mathbb{R}_+$ with the set of all sections of $\mathbb{Q}_+$. For example, the real number which we normally denote by $\sqrt{2}$ appears in this treatment as the section

$$\{s : s \in \mathbb{Q}_+ \text{ and } s^2 < 2\}.$$

We fit $\mathbb{Q}_+$ inside the system $\mathbb{R}_+$ by identifying the rational number r with the section

$$\{s : s \in \mathbb{Q}_+ \text{ and } s < r\}.$$

Equality is defined in $\mathbb{R}_+$ by saying that two sections are equal if and only if they have the same elements (i.e. they are equal as sets).

Addition and multiplication are defined in the following way. It L and M are sections of $\mathbb{Q}_+$, then

$$L + M = \{s + t : s \in L \text{ and } t \in M\}$$
$$LM = \{st : s \in L \text{ and } t \in M\}.$$

An ordering is introduced by writing

$$L \le M \Leftrightarrow L \subset M.$$

In this way, the structure of $\mathbb{R}_+$ is expressed entirely in terms of the structure of $\mathbb{Q}_+$.

Next comes the task of deducing the properties of $\mathbb{R}_+$ from the properties of $\mathbb{Q}_+$ and the definitions given above. We leave the algebraic properties for the next set of exercises and concentrate on the continuum axiom which, as we know from chapter 9, is crucial to the foundations of analysis. In this chapter, of course, we are assuming none of the axioms for the real number system given in previous chapters. Instead we are seeking to justify them. The content of the continuum axiom is therefore a *theorem* in this chapter.

10.14†   *Theorem* Any non-empty subset S of $\mathbb{R}_+$ which is bounded above has a smallest upper bound.

Proof The elements of S are, of course, identified with sections of $\mathbb{Q}_+$. If H is an upper bound for S, then this means that H is a section of $\mathbb{Q}_+$ with the property that, for each section $L \in S$,

$$L \subset H.$$

We define a set B of positive rational numbers by

$$B = \bigcup_{L \in S} L.$$

We shall show that B is a section of $\mathbb{Q}_+$ and hence represents a positive real number and that it is the smallest upper bound of the set S. To show that B is a section we have three things to check.

(i) If $s \in B$ and $r < s$, then $r \in B$. To prove this we observe that, if $s \in B$, then $s \in L$ for some $L \in S$. But L is a section. Hence, if $r < s$, then $r \in L$ and therefore $r \in B$.

(ii) *B* is bounded above. This follows from the fact that, for each $L \in S$, $L \subset H$ and therefore

$$B = \bigcup_{L \in S} L \subset H.$$

(iii) *B* has no maximum. If $m \in B$, then $m \in L$ for some $L \in S$. But $L \subset B$ and so, if *m* is a maximum for *B*, it is a maximum for the section *L*. But *L* has no maximum.

Obviously $L \subset B$ for each $L \in S$. Hence *B* is an upper bound for *S*. Suppose that *B'* is a second upper bound. Then $L \subset B'$ for each $L \in S$ and hence

$$B = \bigcup_{L \in S} L \subset B'.$$

But, by definition, $B \leq B' \Leftrightarrow B \subset B'$. Hence *B* is the *smallest* upper bound.

10.15†   *Exercise*

(1) Let $r \in \mathbb{Q}_+$ and let $L = \{s : s \in \mathbb{Q}_+ \text{ and } s < r\}$. Prove that *L* is a section of $\mathbb{Q}_+$ and that, if this section is identified with the positive rational number *r*, then the definitions given for addition, multiplication and an ordering agree with those for $\mathbb{Q}_+$.

(2) Repeat questions 4, 5 and 6 of exercise 10.12 but with $\mathbb{Q}_+$ replaced by $\mathbb{R}_+$ throughout.

(3) Show that the set $L = \{s : s \in \mathbb{Q}_+ \text{ and } s^2 < 2\}$ is a section of $\mathbb{Q}_+$ and that

$$\{st : s \in L \text{ and } t \in L\} = \{r : r \in \mathbb{Q}_+ \text{ and } r < 2\}.$$

What is the relevance of this result? [*Hint*: See exercise 9.22(4).]

10.16†   **Negative numbers and displacements**

Real numbers are not only used to measure length. We also use them to measure such quantities as time, displacement, speed and acceleration. To motivate the introduction of negative numbers, we shall consider the interpretation in terms of displacement.

Suppose that *x* is a positive real number. We say that an object has undergone a displacement *x* if it has been shifted *x* units to the right. If *a* and *b* are positive real numbers, then the total displacement obtained by following displacement *a* by displacement *b* is simply $a + b$.

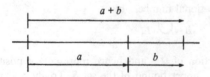

What further displacement *x* is necessary after displacement *b* to produce total displacement *a*?

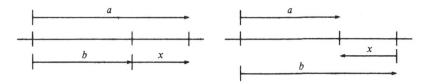

If $b < a$, this question is easily answered since the equation

$$b + x = a$$

then has a unique solution $x \in \mathbb{R}_+$ (exercise 10.15(2)). On the other hand, if $b \geq a$, there is no $x \in \mathbb{R}_+$ which satisfies the equation. Nevertheless, the question still makes good physical sense. One simply needs to take into account displacements to the left as well as to the right. If we are to capture this physical notion in mathematical terms we must *construct* a system in which every equation $b + x = a$ has a solution. This system will be the real number system $\mathbb{R}$.

The objects in $\mathbb{R}$ will represent displacements either to the left or to the right (except for 0 which will represent no shift at all). If $a \in \mathbb{R}$, than $-a$ will represent the displacement in the opposite direction to a by the same number of units.

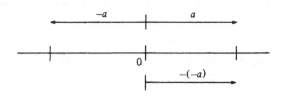

For consistency, it is then necessary that $-(-a) = a$. The rule that 'minus × minus = plus' is therefore not an accidental byproduct of the algebra of the situation (as might appear from chapter 7). It is, on the contrary, an essential feature of the structure of $\mathbb{R}$.

10.17†   Real numbers

The system $\mathbb{R}_+$ of positive real numbers was constructed in §10.13. From $\mathbb{R}_+$ we now construct the system $\mathbb{R}$ of all real numbers.

We wish to construct the system $\mathbb{R}$ in such a way that each equation $b + x = a$ has a solution. As in §10.11, it would be natural to proceed by using the ordered pair (a, b) to stand for the solution. But this approach will not do because, for example, we shall want the equations $b + x = a$ and $b + 1 + x = a + 1$ to have the *same* solution but (a, b) and $(a + 1, b + 1)$ are not the same ordered pair.

We therefore introduce an equivalence relation on $\mathbb{R}_+ \times \mathbb{R}_+$ by writing

$$(a, b) \sim (c, d) \Leftrightarrow a + d = b + c.$$

This equivalence relation splits $\mathbb{R}_+ \times \mathbb{R}_+$ into disjoint equivalence classes (theorem 5.7) and it is these equivalence classes we shall use to represent the elements of $\mathbb{R}$.

It will help our intuition in this context if we use the symbol

$$a - b$$

to label the equivalence class which contains the ordered pair (a, b). We then need to note that

$$a - b = c - d \Leftrightarrow a + d = b + c.$$

If $a > b$, then we know that there is a unique positive real number c with the property $a = b + c$. The positive real numbers may therefore be fitted inside the system by making the identification

$$c = a - b$$

where $a > b$.

Addition and multiplication are introduced into the system by defining

and
$$\left. \begin{array}{l} (a-b)+(c-d)=(a+c)-(b+d) \\ (a-b)(c-d)=(ac+bd)-(bc+ad). \end{array} \right\}$$

With these definitions the system $\mathbb{R}$ is a field (i.e. satisfies all the usual laws of arithmetic). For example, property Iiii (see §7.3) for a field is satisfied by making the identification

$$0 = a - a$$

and property Iiv is satisfied by observing that, given any $a - b \in \mathbb{R}$,

$$(a-b)+(b-a)=(a+b)-(a+b)=0.$$

Thus $-(a-b)$ is simply $(b-a)$ as one would expect.

Finally, we need to introduce an ordering into the field $\mathbb{R}$. Referring to §7.10, we see that this simply amounts to specifying which of the non-zero elements of $\mathbb{R}$ are to be positive and which are to be negative. Obviously we want $\mathbb{R}_+$ to be the set of positive elements of $\mathbb{R}$ and so we make the definition

$$a - b > 0 \Leftrightarrow a > b.$$

Next comes the task of checking that, with these definitions, all the properties of $\mathbb{R}$ may be deduced from the properties of $\mathbb{R}_+$. Fortunately, we know that the properties of $\mathbb{R}$ are summarised by the axioms given in chapters 7 and 9. The task is therefore relatively easy.

10.18†   *Exercise*

(1) Prove that the relation $\sim$ introduced in §10.17 is an equivalence relation on $\mathbb{R}_+ \times \mathbb{R}_+$ – i.e. it satisfies the rules of §5.5.
(2) Prove that the definitions of addition, multiplication and the ordering introduced in §10.17 are consistent with the criterion for equality – i.e. if $x_1 = x_2$ and $y_1 = y_2$ check that

$$x_1 + y_1 = x_2 + y_2$$

$$x_1 y_1 = x_2 y_2$$
$$x_1 > 0 \Leftrightarrow x_2 > 0.$$

(3) Check that the definitions given for addition and multiplication in $\mathbb{R}$ are consistent with those of $\mathbb{R}_+$ in the case when $a > b$ and $c > d$.

(4) Prove that $\mathbb{R}$ is a field (i.e. satisfies axioms I, II and III of §7.3).

(5) Prove that $\mathbb{R}$ is an ordered field (i.e. satisfies axiom IV of §7.10). Check that the ordering on $\mathbb{R}$ is consistent with that of $\mathbb{R}_+$.

(6) Check that $\mathbb{R}$ satisfies the continuum axiom (i.e. every non-empty set S which is bounded above has a smallest upper bound or, equivalently, every non-empty set S which is bounded below has a largest lower bound).

10.19†   Linear and quadratic equations

If a and b are real numbers, then the 'linear equation'

$$ax + b = 0$$

has a unique solution $x \in \mathbb{R}$ provided that $a \neq 0$.

Now consider the 'quadratic equation'

$$ax^2 + bx + c = 0$$

in which a, b and c are real numbers and $a \neq 0$. This may be solved by 'completing the square' provided that $b^2 - 4ac \geq 0$. We have that

$$4a^2 x^2 + 4abx + 4ac = 0$$
$$(2ax + b)^2 - b^2 + 4ac = 0$$
$$(2ax + b)^2 = b^2 - 4ac.$$

Thus

$$2ax + b = \pm \sqrt{(b^2 - 4ac)}$$
$$x = \frac{-b \pm \sqrt{(b^2 - 4ac)}}{2a}.$$

If $b^2 - 4ac > 0$, it follows that the quadratic equation has two *real* roots given by

$$\alpha = \frac{-b + \sqrt{(b^2 - 4ac)}}{2a}, \quad \beta = \frac{-b - \sqrt{(b^2 - 4ac)}}{2a}.$$

It is then easy to check that

$$ax^2 + bx + c = a(x - \alpha)(x - \beta).$$

If $b^2 - 4ac = 0$, we still say that the quadratic equation has two solutions but that these are co-incident. This convention is justified by the fact that, if $b^2 - 4ac = 0$ and α is a root of the quadratic equation, then

$$ax^2 + bx + c = a(x - \alpha)(x - \alpha).$$

What about the case $b^2 - 4ac < 0$? Certainly the quadratic equation will then have no real solutions at all. The graph of $y = ax^2 + bx + c$ will never cut the x-axis.

Evidently the system of real numbers is inadequate to deal with this situation. Just as we had to go beyond the rational number system to find a solution for $x^2 = 2$, so we must go beyond the real number system to find solutions for quadratic equations in which $b^2 - 4ac < 0$.

The number system which we shall construct to deal with this problem is called the complex number system $\mathbb{C}$. This system is a field (i.e. all the usual rules of arithmetic are true for $\mathbb{C}$) which contains an element i satisfying

$$i^2 = -1.$$

Given a quadratic equation $ax^2 + bx + c = 0$ with real coefficients satisfying $b^2 - 4ac < 0$, one can then complete the square as before and so obtain the two *complex* roots

$$\alpha = \frac{-b + i\sqrt{(4ac - b^2)}}{2a},$$

$$\beta = \frac{-b - i\sqrt{(4ac - b^2)}}{2a}.$$

The object i is often regarded by the layman as a highly mysterious entity. For this reason it is said to be an *imaginary* number. Certainly it cannot be a *real* number because no real number has a negative square. Unfortunately, this aura of mystery is quite undeserved as will be apparent from the prosaic nature of the construction given in the next section.

10.20†   Complex numbers

In this section we construct the complex number system $\mathbb{C}$ from the real number system $\mathbb{R}$. We want $\mathbb{C}$ to be a field which contains an element i for which $i^2 = -1$. There is, of course, no point in asking that $\mathbb{C}$ be an *ordered* field. In an ordered field no element has a negative square. We therefore make no attempt to define $z_1 \leqq z_2$ when z_1 and z_2 are complex numbers. In general, the assertion $z_1 \leqq z_2$ is *meaningless* for complex numbers.

After §10.11 and §10.17, it comes as no surprise that we begin with the set $\mathbb{R} \times \mathbb{R}$ of all ordered pairs of real numbers. Each such ordered pair will represent a different

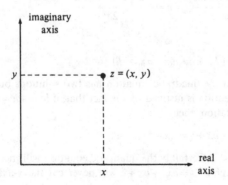

complex number. One can therefore think of a complex number $z = (x, y)$ as a point in the plane.

The real numbers are fitted inside the complex number system by making the identification

$$x = (x, 0).$$

In particular, we have $0 = (0, 0)$ and $1 = (1, 0)$. The complex numbers $(0, y)$ with $y \neq 0$ are said to be *purely imaginary*. Of particular importance is the complex number

$$i = (0, 1).$$

Addition and multiplication of complex numbers are defined by

$$\left.\begin{array}{l} \text{(i)} \;\; z_1 + z_2 = (x_1, y_1) + (x_2, y_2) = (x_1 + x_2, y_1 + y_2) \\[2mm] \text{(ii)} \;\; z_1 z_2 = (x_1, y_1)(x_2, y_2) = (x_1 x_2 - y_1 y_2, x_1 y_2 + y_1 x_2). \end{array}\right\}$$

If z_1 and z_2 are both real (i.e. $y_1 = y_2 = 0$), these formulae reduce to the usual addition and multiplication for real numbers. If α is a real number and z is complex, then (ii) says that

$$\alpha z = (\alpha, 0)(x, y) = (\alpha x, \alpha y).$$

Thus, given any complex number z, we have that

$$z = (x, y) = (x, 0) + (0, y) = x(1, 0) + y(0, 1)$$

– i.e.

$$\boxed{z = x + iy.} \tag{1}$$

This representation leads us to call x the *real part* of z and y the *imaginary part* of z. We write

$$x = \mathcal{R}z : y = \mathcal{I}m\, z.$$

(Note that $\mathcal{I}m\, z \neq iy$.)

Next observe that

$$i^2 = (0, 1)(0, 1) = (0 \cdot 0 - 1 \cdot 1, \; 0 \cdot 1 + 1 \cdot 0) = (-1, 0) = -1$$

– i.e.

$$\boxed{i^2 = -1.} \tag{2}$$

The rules for addition and multiplication of complex numbers are cunningly chosen to ensure that complex numbers satisfy all the usual rules of arithmetic – i.e. $\mathbb{C}$ is a field. (See exercise 10.24(1).) This fact, together with (1) and (2), is all that one needs to remember in order to manipulate complex numbers. In particular, the precise form of the definitions of addition and multiplication of complex numbers can be quietly forgotten.

10.21   Examples

(i) $(1+2i)+(3+4i)=(1+3)+(2+4)i=4+6i.$

(ii) $(1+2i)-(3+4i)=(1-3)+(2-4)i=-2-2i.$

(iii) $(1+2i)(3+4i)=1(3+4i)+2i(3+4i)$
$$=3+4i+6i+8i^2$$
$$=3+4i+6i-8$$
$$=-5+10i.$$

(iv) $\dfrac{(1+2i)}{(3+4i)}=\dfrac{(1+2i)}{(3+4i)}\dfrac{(3-4i)}{(3-4i)}=\dfrac{3-4i+6i-8i^2}{9-12i+12i-16i^2}=\dfrac{11+2i}{25}=\dfrac{11}{25}+\dfrac{2}{25}\,i.$

Note: The trick used for dividing by a complex number in (iv) *always* works. We have that

$$\frac{(a+ib)}{(c+id)}=\frac{(a+ib)(c-id)}{(c+id)(c-id)}=\frac{(a+ib)(c-id)}{c^2+d^2}.$$

10.22†   Cubic equations

After reading the above one might feel that the study of complex numbers is all very well – but why bother? Is there really any point in saying that the roots of $x^2+1=0$ are $x=i$ and $x=-i$? Why not just say that the graph of $y=x^2+1$ never cuts the x-axis?

As an example of a situation where complex numbers are intrinsically useful we consider the problem of finding a formula for the solutions of the cubic equation

$$ax^3+bx^2+cx+d=0$$

where $a\neq0$. This is, of course, an interesting question in itself. The problem was a great challenge to the Italian mathematicians of the sixteenth century but, with modern techniques, it is relatively easy.

First make the substitution $x=Az+B$. The cubic equation then becomes

$$z^3+3Hz+G=0 \tag{1}$$

provided that A and B are chosen so that the coefficient of z^3 is 1 and that of z^2 is 0.

By the binomial theorem,

$$(p+q)^3=p^3+3pq(p+q)+q^3.$$

Hence, if $z=p+q$,

$$z^3-3pqz-(p^3+q^3)=0. \tag{2}$$

Comparing equations (1) and (2), we conclude that

$$G=-(p^3+q^3)$$
$$H=-pq \tag{3}$$
$$H^3=-p^3q^3.$$

But then p^3 and q^3 are the roots of the quadratic equation

$$t^2 + Gt - H^3 = 0.$$

Hence

$$p^3 = \tfrac{1}{2}\{-G \pm (G^2 + 4H^3)^{1/2}\}. \tag{4}$$

Given any p which satisfies (4), the corresponding value of q can be found from (3). The solutions of (1) are therefore given by

$$z = p + q = p - H/p, \tag{5}$$

where p is any complex number satisfying (4).

Consider now the specific example

$$z^3 - 3z = 0.$$

This has been chosen because it is obvious that the roots are $z = 0$, $z = \sqrt{3}$ and $z = -\sqrt{3}$ and hence it will be easy to check our formula. We have that $G = 0$ and $H = -1$. Hence (4) reduces to

$$p^3 = \pm i.$$

We first consider the possibility that $p^3 = i$. It is easy to check that this equation has the three solutions $p = -i$, $p = \tfrac{1}{2}(\sqrt{3} + i)$, $p = -\tfrac{1}{2}(\sqrt{3} - i)$. From $p = -i$ we obtain the solution

$$z = -p - H/p = i - 1/i = -i + i = 0$$

(since $i^2 = -1$, $1/i = -i$). From $p = \tfrac{1}{2}(\sqrt{3} + i)$ we obtain the solution

$$z = \tfrac{1}{2}(\sqrt{3} + i) + \frac{2}{(\sqrt{3} + i)} = \tfrac{1}{2}(\sqrt{3} + i) + \tfrac{2}{4}(\sqrt{3} - i)$$
$$= \sqrt{3}.$$

From $p = -\tfrac{1}{2}(\sqrt{3} - 1)$ we obtain

$$z = -\tfrac{1}{2}(\sqrt{3} - i) - \frac{2}{(\sqrt{3} - i)} = -\tfrac{1}{2}(\sqrt{3} - i) - \tfrac{2}{4}(\sqrt{3} + i)$$
$$= -\sqrt{3}.$$

We have therefore obtained (with much effort) the solutions $z = 0$, $z = \sqrt{3}$ and $z = -\sqrt{3}$ which we knew already. Had we started with $p^3 = -i$ the same solutions would have appeared. The significant thing is that these *real* solutions to a cubic equation with *real* coefficients were obtained via an excursion into the complex number system. Indeed, when a cubic equation with *real* coefficients has three *real* solutions, then $G^2 + 4H^3 < 0$ and hence the formula for the solution of a cubic is useless without a knowledge of complex numbers!

Having dealt with cubic equations, the early Italian mathematicians went on to consider quartics. Eventually a formula for the solutions for quartic equations was found but all attempts at a formula for the roots of quintic equations were unsuccessful. It was not till much later that the great Norwegian mathematician Abel showed that no such formula exists – i.e. that, in general, the roots of a quintic equation, although they certainly exist, cannot be obtained from the coefficients using only a finite number of additions, subtractions, multiplications, divisions and extractions of nth roots.

We take up this point again in the next chapter.

10.23†   Polynomials

A polynomial $P(z)$ is an expression of the form

$$P(z) = a_n z^n + a_{n-1} z^{n-1} + \ldots + a_1 z + a_0.$$

If $a_n \neq 0$, we say that the polynomial is of *degree n*.

A result of vital importance whose proof is beyond the scope of this book is that every polynomial of degree n with complex coefficients has precisely n complex roots in the sense that $P(z)$ can be expressed uniquely in the form

$$P(z) = a_n(z - \alpha_1)(z - \alpha_2) \ldots (z - \alpha_n).$$

It is interesting that an algebraic proof of this algebraic result is not known. Analytic methods must be used.

10.24†   *Exercise*

(1) Prove that $\mathbb{C}$ is a field – i.e. check that axioms I, II and III of §7.3 hold for $\mathbb{C}$. [*Hint*: See examples 10.21.]

(2) Find the roots of the following quadratic equations
   (i) $x^2 - 1 = 0$ (ii) $x^2 + 1 = 0$
   (iii) $3x^2 + 3x + 1 = 0$ (iv) $x^2 + 2x + 1 = 0$.

(3) Calculate the real and imaginary parts of the following complex numbers.

   (i) $(z_1 + z_2)z_3$ (ii) $z_1 z_2 - z_3$ (iii) $\dfrac{z_1 - z_3}{z_2}$ (iv) $\dfrac{z_1 z_3}{z_2}$

   where $z_1 = 1 + 2i$, $z_2 = 2 + 3i$, $z_3 = 3 + 4i$

(4) Prove the following.

   (i) $\dfrac{1}{i} = -i$ (ii) $i^4 = 1$ (iii) $i^3 = -i$ (iv) $\left(\dfrac{1+i}{\sqrt{2}}\right)^2 = i$.

   Find the roots of the quadratic equation $z^2 - i = 0$.

(5) Prove that $p^3 - i^3 = (p - i)(p^2 + ip + i^2)$. [*Hint*: See exercise 8.9(3). Hence find the solutions of $p^3 = -i$. [*Hint*: See question 4iii.] Use these solutions to solve

$$z^3 - 3z = 0$$

   by means of the formula of §10.22.

(6) Let $P(z)$ be a polynomial of degree n. If $P(\zeta) = 0$, show that $P(\zeta) = (z - \zeta)Q(z)$, where $Q(z)$ is a polynomial of degree $n - 1$. [*Hint*: See exercise 8.9(3).]

(7) If $z = x + iy$, its *complex conjugate* is $\bar{z} = x - iy$. Prove that

(i) $\dfrac{}{z_1+z_2}=\dfrac{}{z_1}+\dfrac{}{z_2}$

(ii) $\dfrac{}{z_1 z_2}=\dfrac{}{z_1}\dfrac{}{z_2}.$

(8) If $P(x)=a_n x^n+a_{n-1}.x^{n-1}+\ldots+a_1 x+a_0$ has *real* coefficients (i.e. $a_0, a_1, \ldots, a_n$ are real numbers), prove that, for all complex numbers z

$$P(\bar{z})=\overline{P(z)}.$$

Deduce that ζ is a root of $P(x)$ if and only if $\bar{\zeta}$ is a root of $P(x)$. (See question 7.)

(9) Prove that, if $P(x)$ is a polynomial of *odd* degree with *real coefficients*, then

$$P(x)=0$$

has at least one *real* root. You may assume that a polynomial of degree n has precisely n complex roots in the sense of §10.23. [*Hint:* See question 8.] Give an example of a polynomial of even degree with real coefficients which has no real roots.

11† NUMBER THEORY

11.1† Introduction

In the previous chapter we explained how the real number system may be constructed from the natural numbers in a straightforward and almost inevitable fashion. But the development of new number systems was historically anything but an easy or an automatic process. The introduction of new ideas was often fiercely resisted by the traditionalists. In this, of course, the history of mathematics differs not in the least from the history of anything else. Our legacy from this resistance is the unfortunate and emotive words 'negative', 'irrational', 'imaginary' and the like. In this chapter we shall even meet the surd numbers (other numbers being, of course, 'absurd').

Consider, for example, the word 'irrational'. The philosophers of the school of Pythagoras were familiar, of course, with Pythagoras' theorem. Presumably they based their 'proof' on a diagram like that drawn below and used some argument about the various areas involved which we would nowadays express algebraically by the formula

$$(a+b)^2 = c^2 + 4(\tfrac{1}{2}ab).$$

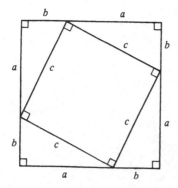
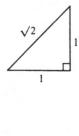

Once Pythagoras' theorem has been discovered, one has little choice but to agree that there is a line segment whose length is $\sqrt{2}$. But, to the Pythagoreans, it was 'obvious' that all lengths are 'commensurable'. This means, in modern terms, that their ratio is rational. When one of the Pythagoreans discovered that $\sqrt{2}$ is irrational, the result was greeted with consternation. An attempt was even made to keep this 'flaw in the universe' a secret. But such a hot item of news was bound to leak out and it did.

Eudoxus later showed how to cope with irrational lengths by observing that they are smaller than all the greater rational lengths and larger than all the smaller rational lengths. This is, of course, essentially the idea we used in constructing $\mathbb{R}_+$ from $\mathbb{Q}_+$ and it seems, with the advantage of hindsight, almost perverse of the Greeks not to have invented the real numbers. But the world had to wait until modern times for Dedekind to make the final step. As we have seen, there is nothing 'irrational' (in the usual sense of the word) about the irrational numbers as constructed by Dedekind and the word we use to describe them is just a picturesque survival of a long dead controversy.

In this chapter, we use the term *'number theory'* to describe the study of the properties of various types of numbers (e.g. primes, surds, irrationals) as opposed to the use to which these numbers are put in other branches of mathematics (e.g. analysis). We therefore regard the theorem (8.15) which asserts that $\sqrt{2}$ is irrational as a result in number theory. Sometimes the term 'number theory' is restricted to the study of properties of the set $\mathbb{N}$ of natural numbers. It is certainly true that the most appealing problems (and probably the hardest) in number theory are those concerned simply with the natural numbers. At first sight, it seems hard to believe that there can be much about the natural numbers that we do not know. We therefore mention three famous unsolved problems about $\mathbb{N}$.

It is shown in Euclid that the set of all prime numbers in $\mathbb{N}$ is infinite. (The proof, incidentally, is a beauty. If $2, 3, 5, \ldots, p$ were the only primes, then $(2 \cdot 3 \cdot 5 \cdot \ldots p) + 1$ would be indivisible by any prime.) But it is quite unknown whether there is an infinite set of primes p such that both p and $p + 2$ are prime. This is called the *'twin prime conjecture'*.

An even more famous unsolved problem is the *'Goldbach conjecture'* which asks: is it true that every even $n > 4$ may be expressed in the form $n = p + q$, where p and q are odd primes (i.e. $p > 2$ and $q > 2$)?

But the most famous of all unsolved problems in number theory is *'Fermat's Last Theorem'*. We begin by asking whether the equation

$$x^2 + y^2 = z^2$$

has a solution (x, y, z) with x, y and z all natural numbers. That $(x, y, z) = (3, 4, 5)$ is a solution was known even to the ancient Egyptians. Fermat, an early French mathematician, posed the same problem for the equation

$$x^n + y^n = z^n$$

where $n \in \mathbb{N}$. After his death, a note was found scribbled in a margin saying that he had discovered a beautiful proof that the equation has no solutions in $\mathbb{N}$ if $n > 2$ but that the margin was too small to write it down. But no trace of a proof was found elsewhere and the truth of the 'theorem' remains unknown. However, Fermat did leave us the proofs of many beautiful theorems in number theory. For example, he showed that a prime is expressible as the sum of two squares if and only if it leaves remainder 1 when divided by 4. Thus

$$1^2 + 2^2 = 5, \quad 2^2 + 3^2 = 13, \quad 1^2 + 4^2 = 17, \quad 2^2 + 5^2 = 29, \ldots$$

but none of the primes $2, 3, 7, 11, 19, 23, \ldots$ can be expressed in this way.

The conjectures described above are far too difficult for us to consider in this book. However, it is ironic to observe that what progress has been made with traditional problems of this kind has largely been with the help of complex analysis and the use of the so-called 'imaginary' number i so fiercely opposed at its inception.

11.2† Integers

The set $\mathbb{Z}$ of *integers* may be defined by

$$\mathbb{Z} = \{m - n : m \in \mathbb{N} \text{ and } n \in \mathbb{N}\}.$$

The elements of $\mathbb{Z}$ are therefore $0, \pm 1, \pm 2, \ldots$ and so on. A positive integer is, of course, the same thing as a natural number.

We begin with the observation that the set $\mathbb{Z}$ is a group under addition. In particular, this means that, if m and n are integers, then so are $m+n$ and $m-n$.

Also, if m and n are integers, then mn is an integer. But m/n need not be an integer (even when $n \neq 0$). Consider the case $m=8$ and $n=3$. In the language of the schoolroom, '3 into 8 won't go' – or, for the more sophisticated, '3 goes into 8 two times with remainder 2'. By this is meant, of course, that

$$8 = (3 \times 2) + 2.$$

The general version of this result is called the division algorithm.

11.3† Division algorithm

If a and b are integers and $b > 0$, then there exist unique integers q and r such that

$$a = bq + r$$

and $0 \leq r < b$ (i.e. 'b goes into a q times with remainder r').

11.4† *Exercise*

(1) Prove that the sum and the difference of two integers is again an integer. Explain why it follows that $\mathbb{Z}$ is a group under addition (see exercise 7.5(6)).
(2) Prove that the product of two integers is again an integer.
(3) Is $\mathbb{Z}$ a group under multiplication?
(4) Prove the division algorithm. [*Hint*: The set $S = \{a - bx : a - bx \geq 0 \text{ and } x \in \mathbb{Z}\}$ is not empty. (Why?) Hence S has a minimum element (exercise 9.18(3)). Call this minimum r and prove that $0 \leq r < b$.]
(5) Use the division algorithm to prove that any integer is either even or else odd, but not both. (See exercise 8.12(3).)
(6) Let S be a non-empty set of integers such that $S \neq \{0\}$. If the sum and the difference of any two elements of S is again an element of S (from which it follows that S is a group under addition), prove that S contains a smallest positive element d and that S consists of all integer multiples of d, i.e.

$$S = \{md : m \in \mathbb{Z}\}.$$

11.5†   Factors

If a and b are integers and 'b goes into a', i.e. $a = bq$, where q is an integer, we write $b|a$ and say that b *divides* a or that b is a *factor* of a.

11.6   *Example* The factors of 6 are ± 1, ± 2, ± 3 and ± 6. The factors of -52 are ± 1, ± 2, ± 4, ± 13, ± 52.

Given two integers a and b, their *greatest common factor* is the largest natural number which divides both of them.

11.7   *Example* The greatest common factor of 6 and -52 is 2. That of 36 and 24 is 12. That of -7 and 12 is 1.

11.8†   Euclid's algorithm

Let a and b be non-zero integers and let d be their greatest common factor. Then integers x and y exist such that

$$d = ax + by.$$

Proof The sum and the difference of any two elements of the set of integers

$$S = \{ax + by : x \in \mathbb{Z} \text{ and } y \in \mathbb{Z}\}$$

obviously also lie in S. From exercise 11.4(6) it follows that S must consist of all integer multiples of the *smallest* $d > 0$ such that

$$d = ax + by \quad (x \in \mathbb{Z} \text{ and } y \in \mathbb{Z}).$$

We show that d is the greatest common factor of a and b. First note that $a = a \cdot 1 + b \cdot 0 \in S$ and $b = a \cdot 0 + b \cdot 1 \in S$. Hence $a = m_1 d$ and $b = m_2 d$ for some integers m_1 and m_2. Thus d is certainly a factor of both a and b. Is it the *greatest* common factor?

Let c be another common factor. Then $c|a$ and $c|b$. Thus $a = q_1 c$ and $b = q_2 c$. Therefore, given any integers x and y, $ax + by = (q_1 x + q_2 y)c$ and so $c|ax + by$. In particular, $c|d$. It follows that d is the greatest common factor of a and b (why?) and this completes the proof.

11.9†   Primes

A natural number p other than 1 which has *no* factors except 1 and p is called a *prime* number.

11.10   *Example* Some primes are 2, 3, 5, 7, 11, 13, 17, 19, 23, 29 and 31. On the other hand, 6, 52 and 1 are not primes.

It is fairly clear that any natural number $n \geq 2$ can be written as the product of prime factors. For example, $52 = 2 \times 2 \times 13$.

What is not so clear is that this factorisation is *unique* (except for the order in which the factors appear). Is it, for example, obvious that

$$19 \times 23 \neq 13 \times 29?$$

11.11†   *Theorem* Let a and b be integers and p a prime. Then $p|ab$ implies $p|a$ or $p|b$.

Proof Suppose that $p|ab$ but that it is not true that $p|a$. Since p is a prime it follows that the greatest common factor of p and a must be 1. From Euclid's algorithm we deduce that

$$1 = xp + ya$$

for some integers x and y. But then

$$b = xpb + yab.$$

Since $p|p$ and $p|ab$, it follows from this equation that $p|b$. This concludes the proof.

11.12†   **Prime factorisation theorem**

Any natural number $n \geq 2$ has a unique factorisation.

Proof We take for granted the easily proved proposition that every natural number can be factorised into primes and concentrate on proving the uniqueness.

Suppose that

$$p_1 \cdots p_n = q_1 \cdots q_k$$

where $n \geq k$ and $p_1, \ldots, p_n$ and $q_1, \ldots, q_k$ are primes. Then $q_k | p_1 \ldots p_n$ and so, by theorem 11.11, $q_k | p_j$ for some j. But p_j is prime and therefore $q_k = p_j$.

Relabelling the primes $p_1 \cdots p_n$, if necessary, we conclude that

$$p_1 \cdots p_{n-1} = q_1 \cdots q_{k-1}$$

and a continuation of the argument above yields the required result.

11.13†   **Rational numbers**

The set $\mathbb{Q}$ of rational numbers is the set of all the fractions

$$\frac{m}{n}$$

where m and n are integers (and $n \neq 0$).

Any rational number may be expressed *uniquely* in the form

$$\frac{m}{n}$$

provided that we insist that n is a natural number. and that m and n have no common factor (except 1). This is a consequence of the prime factorisation theorem.

A rational expressed in this form is said to be in its 'lowest terms'. Thus $\frac{1}{2}$ is in its lowest terms but $\frac{2}{4}$ is not.

11.14   *Example* Show that the equation $x^3 = 2$ can have no rational solution.

Proof Suppose that $x = m/n$ is a rational solution in its lowest terms. Then

$$m^3 = 2n^3.$$

If p is a prime factor of n, then it follows from this equation that $p|m$ (theorem 11.11). But m and n have no common factor (except 1). Thus $n = 1$. We are therefore left with the equation

$$m^3 = 2$$

which cannot hold for any integer m. (Why not?)

11.15†   *Exercise*

(1) Let m and n be natural numbers. Prove that

$$x^2 = \frac{m}{n}$$

has no rational solution unless both m and n are perfect squares (i.e. $m = j^2$ and $n = k^2$ where j and k are natural numbers).

(2) Prove that the equation

$$8x^3 - 6x - 1 = 0$$

has no rational solution.

(3) Suppose that $n \geq 2$ is a natural number and that p is prime. Prove that

$$p^{1/n}$$

is irrational.

(4) Let a and b be rational. Prove that

  (i) $a + b\sqrt{2}$ is irrational unless $b = 0$.
  (ii) $a + b\sqrt{2} = 0 \Leftrightarrow a = b = 0$.
  (iii) If a and b are not both zero, then there exist rational numbers c and d such that

$$(a + b\sqrt{2})^{-1} = c + d\sqrt{2}.$$

(5) Suppose that a and b are real numbers satisfying $a<b$. Prove that there exists an irrational number y such that $a<y<b$. [*Hint*: see question 4i and theorem 9.20.]

(6) Prove that the set

$$\mathcal{H}=\{a+b\sqrt{2} : a\in\mathbb{Q} \text{ and } b\in\mathbb{Q}\}$$

is an ordered field. [*Hint*: See exercise 7.5(6).]

11.16†   Ruler and compass constructions

We know that the set $\mathbb{R}$ of real numbers is a field. The set $\mathbb{R}$ also has many subfields. For example, the set $\mathbb{Q}$ of rational numbers and the set $\mathcal{H}$ of exercise 11.15(6) are subsets of $\mathbb{R}$ which are themselves fields.

Given that $\mathcal{F}$ is a subset of $\mathbb{R}$, how do we decide whether or not $\mathcal{F}$ is a field? Since $\mathbb{R}$ is a field, most of the properties for a field are automatically satisfied by $\mathcal{F}$. All that needs to be checked is that, if x and y are elements of $\mathcal{F}$, then so are $x-y$ and x/y, provided $y\neq 0$. (See exercise 7.5(6).)

Let $\mathcal{F}$ be a subfield of $\mathbb{R}$ and let k be a positive element of $\mathcal{F}$. We know that $\sqrt{k}\in\mathbb{R}$ but we shall suppose that $\sqrt{k}\notin\mathcal{F}$. (For example, if $\mathcal{F}=\mathbb{Q}$ and $k=2$.) The set

$$\mathcal{F}(k)=\{a+b\sqrt{k} : a\in\mathcal{F} \text{ and } b\in\mathcal{F}\}$$

is called a quadratic extension of $\mathcal{F}$. (Thus the set $\mathcal{H}$ of exercise 11.15(6) is the quadratic extension $\mathbb{Q}(2)$ of $\mathbb{Q}$.)

11.17†   *Theorem* Let $\mathcal{F}$ be a subfield of $\mathbb{R}$ and let k be a positive element of $\mathcal{F}$ with the property that $\sqrt{k}\notin\mathcal{F}$. Then the quadratic extension $\mathcal{F}(k)$ of $\mathcal{F}$ is also a field.

Proof It is trivial to prove that, if x and y are elements of $\mathcal{F}(k)$, then so are $x-y$ and xy. The only difficulty lies in showing that, for any $x\in\mathcal{F}(k)$ except $x=0$, it is true that $x^{-1}\in\mathcal{F}(k)$. Certainly $x^{-1}\in\mathbb{R}$ because $\mathbb{R}$ is a field.

Write $x=a+b\sqrt{k}$. Then $a\in\mathcal{F}$ and $b\in\mathcal{F}$ and $x=0$ if and only if $a=b=0$. (Why?) Hence, if $x\neq 0$, then $a^2-b^2k\neq 0$ and also $a-b\sqrt{k}\neq 0$. Thus

$$x^{-1}=\frac{1}{a+b\sqrt{k}}=\frac{a-b\sqrt{k}}{(a+b\sqrt{k})(a-b\sqrt{k})}$$

$$=\left[\frac{a}{a^2-b^2k}\right]-\left[\frac{b}{a^2-b^2k}\right]\sqrt{k}.$$

But $a(a^2-b^2k)^{-1}$ and $-b(a^2-b^2k)^{-1}$ belong to $\mathcal{F}$ because $\mathcal{F}$ is a field. Hence $x^{-1}\in\mathcal{F}(k)$

A *surd* (or to be more precise a *quadratic surd*) is a real number which can be obtained by a finite number of steps from the numbers 0 and 1 using only the operations of addition, multiplication, subtraction and division (except by zero) and the extraction of square roots of positive numbers. Thus the real number

$$x = 1 + \sqrt{\{(\sqrt{2}) + \sqrt{(1 + \sqrt{3})}\}}$$

is a surd.

Suppose that $\mathcal{F}_0$, $\mathcal{F}_1, \ldots,$ $\mathcal{F}_n$ are all subfields of $\mathbb{R}$ and $\mathcal{F}_0 = \mathbb{Q}$ and that each field $\mathcal{F}_k$ is a quadratic extension of the field $\mathcal{F}_{k-1}$ $(k = 1, 2, \ldots, n)$. Then we say that $\mathcal{F}_n$ is a *quadratic field* of order n. It follows from theorem 11.17 that all quadratic surds belong to a quadratic field of some order.

11.18 *Example* The real number

$$x = 1 + \sqrt{\{(\sqrt{2}) + \sqrt{(1 + \sqrt{3})}\}}$$

is a surd of order 4. (Why?)

It is natural to ask: are all irrational numbers quadratic surds? Or, if you like: are any irrational numbers 'absurd'?

This is a question which, though they did not know it, would have been of great interest to the classical Greek geometers. A central issue in classical geometry was the invention of 'ruler and compass' constructions. For a ruler and compass construction one is provided with an *unmarked* ruler and a pair of compasses which collapses when it is lifted from the paper after drawing a circle. Thus it cannot be used as a pair of dividers for measuring purposes but only for drawing a circle with a given centre through a given point. The Greeks set themselves the following problems for solution using only ruler and compasses.

(1) Given a circle, construct a square of equal area ('squaring the circle').
(2) Given an angle, construct two lines which trisect the angle ('trisecting the angle').
(3) Given a cube, construct another cube of twice the volume ('duplicating the cube').

The Greeks were quite unable to solve these problems, which is not surprising because they are all *impossible*. That is to say, no such ruler and compass construction can exist. To show that problems (2) and (3) are impossible is not hard. Problem (1) however is not so simple and we are unable to discuss it here.

Consider the 'duplication of the cube'. We begin with two points and take the distance between them as our unit of length. We then have to construct, using only ruler and compasses, a line segment of length $2^{1/3}$.

Unlike the Greeks we can appeal to co-ordinate geometry. We take our original points to be $(0, 0)$ and $(1, 0)$. Any further points that we construct using ruler and compasses will have co-ordinates which can be obtained by solving two simultaneous equations each of which has one of the forms

$$Ax + By + C = 0 \quad \text{or} \quad x^2 + y^2 + Dx + Ey + F = 0$$

where the coefficients A, B, C, D, E and F depend in a simple way on the co-ordinates of the points already constructed.

The important observation to make from this discussion is that the co-ordinates of any points we may construct must be obtained from 0 and 1 by a finite number of applications of the operations of addition, multiplication, subtraction, division (except by zero) and the extraction of square roots of positive numbers. Thus the only points we can construct will be those whose co-ordinates are *quadratic surds*. Thus, for the 'duplication of the cube' to be possible, the real number $2^{1/3}$ must be a quadratic surd.

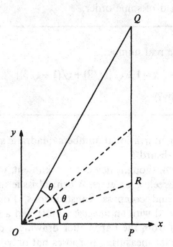

To 'trisect the angle' one is required, given an angle, to construct another angle of one third the magnitude. Start with an angle of 60° as in the diagram. If OP is taken as the unit of length and the trisection of the angle is possible using ruler and compasses, then the point R must have co-ordinates which are quadratic surds. In particular, $\cos \theta$ will be a quadratic surd.

But $\cos 60° = \frac{1}{2}$ and hence

$$\tfrac{1}{2} = \cos 3\theta = 4 \cos^3\theta - 3 \cos \theta$$

and thus the equation

$$8x^3 - 6x - 1 = 0$$

must have a solution which is a quadratic surd given that the trisection of the angle is possible.

The problems of the possibility of 'duplicating the cube' and 'trisecting the angle' therefore reduce to asking whether the equations $x^3 = 2$ and $8x^3 - 6x - 1$ have a solution which is a quadratic surd.

11.19†   *Exercise*

(1) Let a, b and c be elements of a subfield $\mathcal{F}$ of the real number system. Suppose that $c > 0$ and that $\sqrt{c} \notin \mathcal{F}$. Prove that

$$a + b\sqrt{c} = 0 \Leftrightarrow a = b = 0.$$

(2) With the notation of the previous question, the *conjugate* of the element $x = a + b\sqrt{c}$ is defined by $\tilde{x} = a - b\sqrt{c}$. Prove the following.

   (i) $x = 0 \Leftrightarrow \tilde{x} = 0$
   (ii) $x\tilde{x} = a^2 - cb^2$
   (iii) $x = \tilde{x} \Leftrightarrow x = a$
   (iv) $z = x + y \Rightarrow \tilde{z} = \tilde{x} + \tilde{y}$
   (v) $z = xy \Rightarrow \tilde{z} = \tilde{x}\tilde{y}$
   (vi) $z = x^n \Rightarrow \tilde{z} = (\tilde{x})^n$

(3) Suppose that P is a polynomial of degree $n > 0$ whose coefficients all belong to the field $\mathcal{F}$. If $x \in \mathcal{F}(c)$ prove that

$$P(\tilde{x}) = \widetilde{P(x)}.$$

Deduce that $x \in \mathcal{F}(c)$ is a solution of the polynomial equation $P(x) = 0$ if and only if $\tilde{x}$ is a solution.

This result is obvious in the case of a *quadratic* polynomial from the formula for its roots. (Why?)

(4) Suppose that the cubic polynomial $x^3 + Ax^2 + Bx + C$ has two real roots x_1 and x_2. Show that, for some x_3,

$$x^3 + Ax^2 + Bx + C = (x - x_1)(x - x_2)(x - x_3)$$

for all x and that $x_1 + x_2 + x_3 = -A$.

If A, B and C all belong to a field $\mathcal{F}$ of real numbers and the cubic polynomial $x^3 + Ax^2 + Bx + C$ has a root in the quadratic extension $\mathcal{F}(k)$, prove that it must also have a root in $\mathcal{F}$.

(5) Show that a cubic polynomial with rational coefficients cannot have a quadratic surd root unless it also has a rational root. Deduce that the equations

$$x^3 = 2$$

and

$$8x^3 - 6x - 1 = 0$$

have no quadratic surd solutions and hence that the 'duplication of the cube' and the 'trisection of the angle' are impossible.

11.20† Radicals

We have seen that not all real numbers are quadratic surds. Given this result, it is natural to ask, are all real numbers radicals? A *radical* is a number which can be obtained from 0 and 1 by the use of a finite number of the operations of addition, multiplication, subtraction and division (except by zero) and the extraction of nth roots of positive numbers.

The answer to this question is no. Only in exceptional cases is a root of a polynomial of fifth degree or higher with rational coefficients a radical. Formulae are known for the roots of polynomials of degree four or less which express the roots in terms of the coefficients of the polynomial and a finite number of the operations of addition, multiplication, subtraction and division (except by zero) and the extraction of nth roots. But *no* such formula exists for quintic polynomials. This is a famous result of the great Norwegian mathematician Abel.

11.21† Transcendental numbers

It is not even true that every real number is the solution of a polynomial equation

$$a_n x^n + a_{n-1} x^{n-1} + \dots a_0 = 0$$

in which the coefficients $a_0, a_1, \dots, a_n$ are rational numbers. This is proved in the next chapter.

Those real numbers which are the solution of such an equation are called *algebraic*. Those which are not are called *transcendental*. Examples of transcendental numbers are e and π, though it is very hard to *prove* that they are transcendental.

Note, in particular, that since π is transcendental it cannot be a quadratic surd and hence the 'squaring of the circle' is impossible. This result is due to the German mathematician, Lindemann.

12 CARDINALITY

12.1 Counting

How do we decide how many elements there are in a set? Or, to be more prosaic, how would we go about calculating the number of cows in a field?

The answer is that we would count them. But of what does counting consist? Roughly speaking, what we do is this. We look at one of the cows and think 'one', then we look at a different cow and think 'two' and so on until all the cows have been assigned a natural number. We must be careful, of course, to make sure that *all* of the cows have been counted and that none of them has been counted twice. If the cows are frisky and refuse to stand still we might even go so far as to paint a number on their sides to make sure of this.

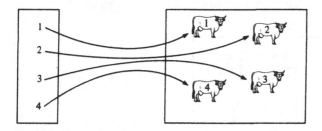

The picture is supposed to represent the process of counting four cows in a field. It is clear that this amounts to setting up a 1 : 1 correspondence between the set $\{1, 2, 3, 4\}$ and the set S of cows in the field, i.e. we construct a bijection $f: \{1, 2, 3, 4\} \to S$.

This leads to the following definition. We shall say that a set S contains n elements if there exists a bijection

$$f: \{1, 2, 3, \ldots, n\} \to S.$$

If, for some $n \in \mathbb{N}$, a set S has n elements (or $S = \emptyset$), we shall say that S is a *finite* set. Otherwise S is an *infinite* set.

(*Note*: Do not confuse the word 'finite' with the word 'bounded'. Every finite set of real numbers is bounded but the interval [0, 1] is an example of an infinite bounded set of real numbers.)

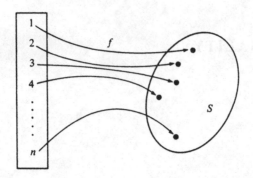

12.2   Cardinality

It certainly makes sense to talk about the number of elements in a finite set. But does it make any sense to discuss the number of elements in an infinite set? The answer to this question is yes – provided that we look at things in the right way.

The vital idea is that of cardinality. We say that two sets A and B have the same *cardinality* (or, if you like, the same 'number of elements') if there exists a bijection $f: A \to B$. The standard example is that of a hall in which all the people are sitting down and no chairs are left unoccupied.

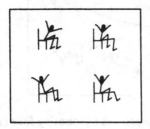

This establishes for us a 1:1 correspondence between the set of all people in the hall and the set of all chairs in the hall.

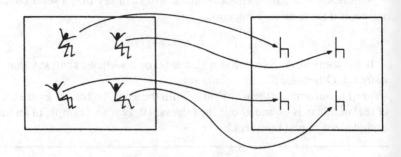

We can therefore say straight away that there are the same number of people as there are chairs in the hall – i.e. that the set of people has the same cardinality as the set of chairs.

Notice that we arrive at this conclusion without having to count at all.

Of course, infinite sets are not like finite sets and we would get into terrible trouble if we treated them as though they were. Consider for example, the following situation.

Let $\mathcal{E}$ denote the set of all *even* natural numbers and consider the function $f : \mathbb{N} \to \mathcal{E}$ defined by

$$f(n) = 2n.$$

Since f is a bijection, the sets $\mathbb{N}$ and $\mathcal{E}$ have the *same* cardinality even though $\mathcal{E}$ is a proper subset of $\mathbb{N}$.

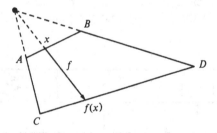

This is a version of Galileo's paradox. Another version of the same paradox is the following. We consider two line segments in the plane and set up a 1 : 1 correspondence between them as illustrated in the diagram.

Then the line segment AB is seen to have the same cardinality as CD. Galileo found this hard to understand on the grounds that the longer segment should have 'more' elements.

Nowadays we do not let such considerations worry us. We simply accept that infinite sets do not behave like finite sets.

12.3   *Exercise*

(1) *Prove* the mapping $f : \mathbb{N} \to \mathcal{E}$ defined by $f(n) = 2n$ is a bijection. Show that

the set $\mathbb{N}$ and the set of all perfect squares (i.e. 1, 4, 9, 16. . . .) have the same cardinality.

(2) If $a < b$, prove that the function $f:(a, b) \to (c, d)$ defined by

$$f(x) = \left(\frac{d-c}{b-a}\right)(x-a) + c$$

is a bijection from the interval (a, b) to the interval (c, d). Deduce that these intervals have the same cardinality.

(3) Show that the set $\mathbb{R}$ and the set $(0, \infty)$ both have the same cardinality as the interval $(0, 1)$.

(4) Let $\mathcal{W}$ be a collection of sets and let $A \in \mathcal{W}$ and $B \in \mathcal{W}$. Write $A \sim B$ if and only if A and B have the same cardinality. Prove that $\sim$ is an equivalence relation on $\mathcal{W}$ and, in particular, that '$A \sim B$ and $B \sim C$ implies $A \sim C$.

(5) Suppose that $f : \{1, 2, \ldots, m\} + S$ and $g : \{1, 2, \ldots, n\} \to T$ are bijections. If $S \cap T = \emptyset$, construct a bijection $h : \{1, 2, \ldots, m+n\} \to S \cup T$. Deduce that the union of two finite sets A and B is finite. [*Hint:* $A \cup B = A \cup (B \setminus A)$.] Hence prove that, if C is infinite and D is finite, then $C \setminus D$ is infinite. Deduce that any subset of a finite set is finite. [*Hint:* suppose that $C \subset D$.]

(6) Show that the union of a finite set and an infinite set is infinite.

12.4   Countable sets

A set S is *countable* if it is finite or has the same cardinality as $\mathbb{N}$. Thus a set S is countable if and only if its elements can be 'counted' – i.e. a distinct natural number can be assigned to each of its elements. In the case of an infinite countable set, this amounts to the assertion that the elements of S can be arranged in a *sequence* $x_1, x_2, x_3, \ldots$.

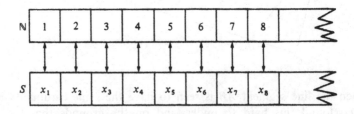

12.5   *Theorem* Any subset M of $\mathbb{N}$ is countable.

Proof We assume that M is not finite and construct a bijection $f : \mathbb{N} \to M$. This is defined inductively by $f(n) = m_n$ where

   (a) $m_1 = \min(M)$ (b) $m_{n+1} = \min(M \setminus \{m_1 m_2, \ldots, m_n\})$.

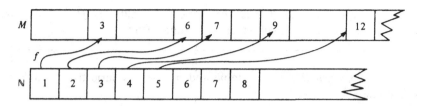

12.6   *Theorem* Any subset T of a countable set S is countable.

   Proof If S is finite, then T is finite by exercise 12.3(5). If S is infinite, there exists a bijection $f: S \to \mathbb{N}$. It follows that T has the same cardinality as $f(T)$. But $f(T)$ is a subset of $\mathbb{N}$ and hence is countable by theorem 12.5.

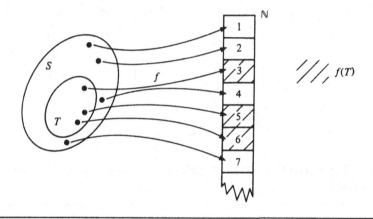

   A simple consequence of the definition is that, for each countable set S, there exists an *injection* $f: S \to \mathbb{N}$ and a *surjection* $g: \mathbb{N} \to S$. (See exercise 12.13(1).) The next two theorems provide converses of these results.

12.7   *Theorem* If S is countable and $F: T \to S$ is injective, then T is countable.

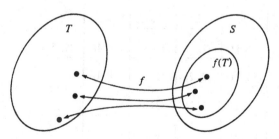

Proof The set T has the same cardinality as $f(T)$ which is countable because it is a subset of the countable set S.

12.8   Theorem If S is countable and $g: S \to T$ is surjective, then T is countable.

Proof Define a function $f: T \to S$ so that $f(t) \in g^{-1}(\{t\})$. Then f is injective, because $t \neq u$ implies $g^{-1}(\{t\}) \cap g^{-1}(\{u\}) = \emptyset$. It follows from theorem 12.7 that T is countable.

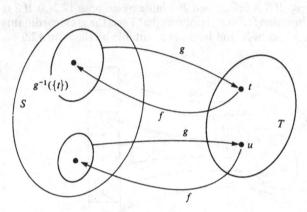

(Recall §6.8. The function f is a 'right-inverse' of g – i.e. $g \circ f$ is the identity function.)

12.9   Example The set $\mathbb{N}^2 = \mathbb{N} \times \mathbb{N}$ is countable. This is a somewhat surprising result at first sight since $\mathbb{N} \times \mathbb{N}$ seems to be a much 'larger' set than $\mathbb{N}$. But we already know from Galileo's paradox that our intuition in this area (which is based entirely on our experience of *finite* sets) is not to be trusted where infinite sets are concerned.

Proof Each ordered pair (m, n) in $\mathbb{N} \times \mathbb{N}$ may be thought of as specifying a square in the array below.

	1	2	3	4	5
1	(1, 1)	(1, 2)	(1, 3)	(1, 4)	(1, 5)
2	(2, 1)	(2, 2)	(2, 3)	(2, 4)	
3	(3, 1)	(3, 2)	(3, 3)		
4	(4, 1)				

We can then establish a bijection between these squares and the natural members by 'counting' them in the fashion illustrated below.

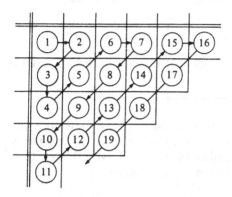

It is not particularly hard to write down a formula for the function $f: \mathbb{N} \times \mathbb{N} \to \mathbb{N}$ defined by this process and then to check that f is indeed a bijection from $\mathbb{N}^2$ to $\mathbb{N}$. However, it is even easier to see that the function $g: \mathbb{N} \times \mathbb{N} \to \mathbb{N}$ defined by

$$g(m, n) = 2^m \, 3^n$$

is an *injection*. Observe that $g(m, n) = g(M, N)$ implies $(m, n) = (M, N)$ by the prime factorisation theorem. The fact that $\mathbb{N} \times \mathbb{N}$ is countable then follows from theorem 12.7.

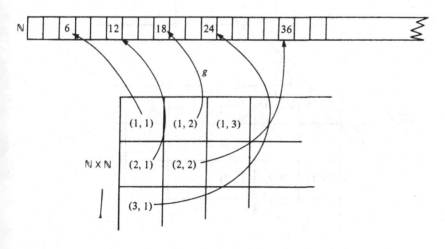

The next theorem generalises the result discussed in the previous example.

12.10   Theorem If S and T are countable, then so is $S \times T$.

Proof Let $f: S \to \mathbb{N}$ and $g: T \to \mathbb{N}$ be injective and suppose that $h:$ $S \times T \to \mathbb{N} \times \mathbb{N}$ is defined by

$$h(s, t) = (f(s), g(t)).$$

Then h is injective and so the result follows from theorem 12.7 and example 12.9.

12.11   Example The set $\mathbb{Q}_+$ of all positive rational numbers is countable. Again this is a surprising result. The set $\mathbb{Q}_+$ and the set $\mathbb{N}$ have the same cardinality (or the same 'number of elements') even though $\mathbb{Q}_+$ has so 'many' elements that it is dense in $(0, \infty)$. (See §9.19.)

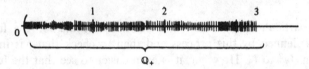

Proof The function $f: \mathbb{N} \times \mathbb{N} \to \mathbb{Q}_+$ defined by

$$f(m, n) = m/n$$

is a surjection. This follows immediately from the definition of $\mathbb{Q}_+$. Hence $\mathbb{Q}_+$ is countable by theorem 12.8 and example 12.9.

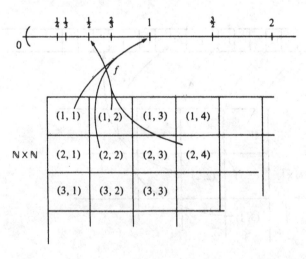

12.12   Theorem The union T of any countable collection $\mathcal{W}$ of countable sets is countable.

Proof Let $F: \mathbb{N} \to \mathcal{W}$ be a surjection. Then $F(n)$ is a set in the collection $\mathcal{W}$ and hence is countable. It follows that there exists a surjection $f_n: \mathbb{N} \to F(n)$.

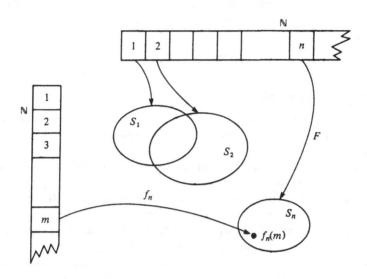

The function $G: \mathbb{N} \times \mathbb{N} \to T$ defined by

$$G(m, n) = f_n(m)$$

is therefore a surjection. (It chooses the 'mth object' in the 'nth set'.) The fact that T is countable then follows from theorem 12.8.

12.13   Exercise

(1) Prove that, if S is countable, then there exists an injective function $f: S \to \mathbb{N}$ and a surjective function $g: \mathbb{N} \to S$.

(2) Prove that the Cartesian product
$$A_1 \times A_2 \times \ldots \times A_n = \{(a_1, a_2, \ldots, a_n): a_1 \in A_1, a_2 \in A_2, \ldots, a_n \in A_n\}$$
of a *finite* number of countable sets $A_1, A_2, \ldots, A_n$ is countable. [*Hint*: Write $A_1 \times A_2 \times \ldots \times A_n = B \times A_n$ where $B = A_1 \times A_2 \times \ldots \times A_{n-1}$ and argue by induction.]

(3) Prove that the set $\mathbb{Z}$ of all integers is countable and also that the set $\mathbb{Q}$ of all rational numbers is countable. Deduce that $\mathbb{Q}^n$ is countable for any natural number n.

(4) Prove that the set of all polynomials of degree n with *rational* coefficients is countable. Deduce that the set of *all* polynomials with rational coefficients is countable. [*Hint*: See theorem 12.12.]

(5) Suppose that S is a countable infinite set and that $g: \mathbb{N} \to S$ is a bijection.

Let $\xi = g(1)$. Prove that the function $f: S \to S \setminus \{\xi\}$ defined by $f(x) = g(g^{-1}(x) + 1)$ is a bijection.

(6) Explain why any infinite set S has a countable infinite subset T. Hence show that any infinite set A contains a proper subset B (i.e. $B \subset A$ but $B \neq A$) such that A and B have the same cardinality. [*Hint*: Use question (5).]

12.14   Uncountable sets

An infinite set S is countable if and only if there exists an injection $f: S \to \mathbb{N}$. But suppose such a function does *not* exist. Then we say that S is *uncountable*.

Roughly speaking, an uncountable set has 'so many' elements that they cannot be arranged in a sequence.

One might be forgiven for doubting the existence of such sets, especially in view of the last section. We saw there that the set $\mathbb{Q}$ of rational numbers, although it seems 'much larger' than $\mathbb{N}$, has the same cardinality as $\mathbb{N}$. If this is true of $\mathbb{Q}$, why not of all infinite sets?

12.15   *Theorem* Let S be any non-empty set. Then the cardinality of the set $\mathscr{P}(S)$ of all subsets of S differs from that of S.

Proof Let $f: S \to \mathscr{P}(S)$. Then, for each $x \in S$, $f(x)$ is a subset of S and so

$$T = \{x : x \in S \text{ and } x \notin f(x)\}$$

defines T as a subset of S. If $x \in S$, then

$$x \in T \Leftrightarrow x \notin f(x).$$

If there exists a $t \in S$ such that $T = f(t)$, we therefore obtain a contradiction by taking $x = t$. Hence f cannot be surjective and the theorem follows.

The above proof is reminiscent of Russell's paradox and the barber who shaves everybody who does not shave himself (§4.13).

When applied to a set S containing n elements, the theorem provides the unsurprising information that $n \neq 2^n$. This is because of a set with n elements has precisely 2^n subsets (including $\emptyset$ and S itself).

A much more interesting result is obtained by taking $S = \mathbb{N}$. We obtain that $\mathscr{P}(\mathbb{N})$, the set of all sets of natural numbers, does *not* have the same cardinality as $\mathbb{N}$. Since $\mathscr{P}(\mathbb{N})$ is clearly not finite, it follows that $\mathscr{P}(\mathbb{N})$ is *uncountable*.

Thus uncountable sets exist. What is more, the theorem tells us that there is no set S of 'greatest cardinality'. Whatever set S may be considered, one can always find another set T which has 'too many' elements to be enumerated using the elements of S.

12.16   *Theorem* If S is uncountable and $S \subset T$, then T is uncountable.

Proof If T is countable, there exists an injection $f\colon T \to \mathbb{N}$. But the restriction of f to S is then an injection from S to $\mathbb{N}$ and this is a contradiction.

12.17†   Decimal expansions

The set $\mathbb{R}$ of all real numbers is an important example of an uncountable set. The easiest proof of this involves 'decimal expansions' which we discuss briefly below.

If all the numbers a_k ($k = 1, 2, \ldots$) are chosen from the set $\{0, 1, 2, \ldots, 9\}$, then the series

$$\frac{a_1}{10} + \frac{a_2}{10^2} + \frac{a_3}{10^3} + \frac{a_4}{10^4} + \ldots$$

converges (by comparison with a geometric progression) to a sum x in the interval $[0, 1]$. We write

$$x = 0.a_1 a_2 a_3 \ldots$$

and say that the right-hand side is a *decimal expansion* of x.

12.18†   *Proposition* Every real number x in the interval $[0, 1]$ has a unique decimal expansion unless it is a rational number of the form $m/10^n$ in which case it has precisely *two* decimal expansions.

We indicate an argument to show the existence of decimal expansions for each $x \in [0, 1]$. The argument is by induction.

Suppose that $a_1, a_2, \ldots, a_n$ have been chosen from $\{0, 1, 2, \ldots, 9\}$ in such a way that

$$e_n = x - \left[\frac{a_1}{10} + \frac{a_2}{10^2} + \ldots + \frac{a_n}{10^n} \right]$$

satisfies

$$0 \leqq e_n < \frac{1}{10^n}. \tag{1}$$

Let a_{n+1} be chosen from $\{0, 1, 2, \ldots, 9\}$ so that a_{n+1} is maximised subject to the constraint $e_{n+1} \geqq 0$.

If $a_{n+1} < 9$, then

$$e_{n+1} < \frac{1}{10^{n+1}}$$

because otherwise we could replace a_{n+1} by $a_{n+1} + 1$. If $a_{n+1} = 9$, things are even simpler since then

$$e_{n+1} = e_n - \frac{9}{10^{n+1}} < \frac{1}{10^n} - \frac{9}{10^{n+1}} = \frac{1}{10^{n+1}}.$$

From (1) and the 'sandwich theorem for convergent sequences' we may deduce that $e_n \to 0$ as $n \to \infty$ and hence that

$$x = \sum_{n=1}^{\infty} a_n 10^{-n}.$$

We leave the issue of uniqueness to the reader except for the observation that, where *two* distinct expansions exist $(x = m10^{-n})$, one of them consists of zeros from some point on and the other consists of nines from some point on. For example,

$$\frac{1}{2} = \frac{5}{10} = .5000 \ldots = .49999 \ldots .$$

12.19   *Theorem* The set $\mathbb{R}$ is uncountable.

Proof Let S be the subset of $\mathbb{R}$ consisting of those $x \in [0, 1]$ which have a unique decimal expansion. We show that S is uncountable. The fact that $\mathbb{R}$ is uncountable then follows from theorem 12.16.

Consider a function $f \colon \mathbb{N} \to S$. Write $x_n = f(n)$ and let the decimal expansion of x_n be

$$x_n = .a_{n1}a_{n2}a_{n3}a_{n4} \ldots .$$

Let $y \in [0, 1]$ be the real number with decimal expansion

$$y = .b_1 b_2 b_3 b_4 \ldots$$

where b_n is defined by

$$b_n = \begin{cases} 8, & a_{nn} \neq 8 \\ 1, & a_{nn} = 8. \end{cases}$$

Then $y \in S$ (because its decimal expansion contains no 0 or 9).

For *no* value of $n \in \mathbb{N}$ is it true that $y = f(n)$. The reason is that, for each $n \in \mathbb{N}$, the decimal expansion of y differs from that of x_n in its nth place. We conclude that no function $f \colon \mathbb{N} \to S$ can be surjective. Hence S is uncountable.

The diagram below illustrates how y is constructed given some specific values of $x_1, x_2, x_3, \ldots,$

$y =$	.	8	8	8	1	8	$\ldots$
$x_1 =$	.	③	1	4	1	5	9 $\ldots$
$x_2 =$	.	1	④	1	4	2	1 $\ldots$
$x_3 =$	.	3	3	③	3	3	3 $\ldots$
$x_4 =$	.	1	4	2	⑧	5	7 $\ldots$
$x_5 =$	.	1	7	3	2	⓪	5 $\ldots$

Observe that y differs from x_1 in its first decimal place and from x_2 in its second decimal place and so on.

12.20†   Transcendental numbers

In §11.21 it was explained that an algebraic number x is one which satisfies a polynomial equation

$$a_n x^n + a_{n-1} x^{n-1} + \ldots + a_1 x + a_0 = 0$$

in which the coefficients $a_0, a_1, \ldots, a_n$ are *rational numbers*.

A real number which is not algebraic is called transcendental. It is exceedingly difficult to prove that any given real number (such as e or π) is transcendental but the following theorem, nevertheless, tells us that 'nearly all' real numbers are transcendental.

The theorem is interesting in that it provides a non-constructive proof of the existence of transcendental numbers (see example 3.15). Such a proof is, of course, less intuitively satisfying than a constructive proof and when Cantor, the originator of the theory described in this chapter, first proved the theorem it was received in a rather hostile fashion. This was doubtless exacerbated by the fact that its proof is very easy compared with Liouville's constructive proof which appeared at around the same time.

12.21†   *Theorem* The set $\mathcal{A}$ of algebraic numbers is countable and hence the set $\mathbb{R} \backslash \mathcal{A}$ of transcendental numbers is uncountable

$\vdots$

Proof Each polynomial equation has at most n roots. It follows that $\mathcal{A}$ is the union of a collection $\mathcal{W}$ of *finite* sets. The collection $\mathcal{W}$ contains one set for each polynomial with rational coefficients. By exercise 12.13(4) it follows that $\mathcal{W}$ is countable. Hence $\mathcal{A}$ is countable by theorem 12.12.

Since $\mathbb{R} = \mathcal{A} \cup (\mathbb{R} \backslash \mathcal{A})$ it follows that $\mathbb{R} \backslash \mathcal{A}$ is uncountable. Otherwise $\mathbb{R}$ would be the union of two countable sets and hence itself countable.

12.22† *Exercise*

(1) Prove that there are precisely 2^n subsets of $\{1, 2, 3, \ldots, n\}$.

(2) It is usual to denote the set of all functions $f: A \to \{0, 1\}$ by 2^A. Show that the mapping $F: 2^A \to \mathscr{P}(A)$ defined by

$$F(f) = f^{-1}(\{1\})$$

is a bijection. Deduce that 2^N is uncountable.

Explain why the set of *all* sequences of 0s and 1s is uncountable.

(3) Explain why the cardinality of $\mathbb{R}$ is *not* the same as the cardinality of the set of all functions $f: \mathbb{R} \to \mathbb{R}$.

(4) If $A_1, A_2, A_3, \ldots$ is a sequence of sets, the set $B = A_1 \times A_2 \times A_3 \times \ldots$ can be identified with the set of all sequences $\langle a_n \rangle$ with the property that $a_n \in A_n$ $(n = 1, 2, \ldots)$.

Give an example to show that B need not be countable although each set A_n is countable. [*Hint*: Use exercise 12.22(2).]

(5) Show that $\mathbb{R}^2 = \mathbb{R} \times \mathbb{R}$ has the same cardinality as $\mathbb{R}$.

(6) If S is countable and T is uncountable, show that $S \cup T$ has the same cardinality as T. [*Hint*: Use exercise 12.13(5) and (6).]

12.23† Counting the uncountable

As we have seen, certain sets are uncountable. If we wish to say more about such sets we must try and think of some way of 'counting' them. Obviously there is no point in trying to 'count' them with the symbols 1, 2, 3, …. If we could, the set we were trying to 'count' would be countable.

To 'count' an uncountable set we need more symbols than are available in the set $\mathbb{N}$ of natural numbers. What should these new symbols be and what properties should they have? We begin by considering more closely the process of counting in the cases we already understand.

Suppose that S is a countable set. Then the act of counting S is essentially the same thing as arranging the elements of S in an *order*.

For example, suppose that we count the set $\{3, 7, 22, 43\}$ as suggested below.

$$
\begin{array}{cccc}
3 & 7 & 22 & 43 \\
\updownarrow & \updownarrow & \updownarrow & \updownarrow \\
4 & 2 & 1 & 3
\end{array}
$$

This will impose the ordering specified by

$$22 \vartriangleleft 7 \vartriangleleft 43 \vartriangleleft 3.$$

We use the notation $\vartriangleleft$ rather than $\leq$ to emphasise that the ordering has nothing at all to do with the magnitude of the objects involved: it simply reflects the order in which they were counted.

Suppose therefore that the act of counting the set S has imposed upon S the ordering $\trianglelefteq$. What properties does this ordering have? The ordering is certainly a total ordering of S in the sense of §5.8. But from the view we are pursuing here, the significant thing is that $\trianglelefteq$ defines a 'well-ordering' of S.

A set S is said to be *well-ordered* by $\trianglelefteq$ if the ordering is a total ordering with the property that each subset T of S has a first element relative to the given ordering.

For example, consider the set $\mathbb{N}$ of natural numbers with the natural ordering $\leq$ by increasing magnitude – i.e.

$$1 < 2 < 3 < \dots.$$

The proof that $\leq$ is a well-ordering was the content of exercise 9.10(6).

For another example of a well-ordering of $\mathbb{N}$, one can imagine an eccentric aesthete who regards odd numbers as more 'beautiful' than even numbers. But, given this distinction, the smaller the number the more 'beautiful' he feels it is. If asked to list the natural numbers in increasing order of 'ugliness' he would write

$$1 \triangleleft 3 \triangleleft 5 \triangleleft \dots \triangleleft 2 \triangleleft 4 \triangleleft 6 \triangleleft \dots$$

where one must imagine all the odd numbers to be listed before 2 is written down. Note that, in this well-ordering, 2 has no immediate predecessor.

A much more complicated well-ordering of the set $\mathbb{N}$ is that obtained by arranging the natural numbers in alphabetical order – i.e. eight, eighteen, eight hundred,... .

The reader may find it instructive to examine the structure of this ordering and to ask himself why it is a well-ordering.

Of course, not all total orderings are well-orderings. For example, the ordering $\leq$ of the positive rational numbers by increasing magnitude is *not* a well-ordering. The set of all positive rationals r such that $r^2 > 2$ has no first element.

We have introduced the notion of a well-ordering so that we can discuss the *well-ordering principle*. This asserts that *any* set can be well-ordered.

Cantor regarded this principle as more or less obvious because of the following argument. Given a set S, pick out one of its elements and call it the first element. Now pick one of the remaining elements and call it the second element. Now suppose that S has been split into two sets A and B by this process, where A is well-ordered and B still has to be dealt with. We then simply pick out one of the elements of B and call it the first element which follows all of the elements of A. In this way we continue until all the elements of S have been exhausted.

It is important to understand that we do *not* assume that S is finite. If S is infinite our process of picking out elements one by one from S will eventually lead to a well-ordered subset A containing an infinite number of elements

$$x_1 \triangleleft x_2 \triangleleft x_3 \triangleleft \dots.$$

We then simply choose an element y from the remaining set $B = S \setminus A$ and write

$$x_1 \triangleleft x_2 \triangleleft x_3 \triangleleft \dots \triangleleft y.$$

The element y then has no immediate predecessor – but then, neither did 2 in the well-ordering $1 \triangleleft 3 \triangleleft 5 \triangleleft \dots \triangleleft 2 \triangleleft 4 \dots$ of the natural numbers that we discussed above.

Cantor's argument is undeniably very compelling. On the other hand, the reader may feel a little unease at the god-like manner in which we have to assume an infinite number of things have been done before we can go on to the next step.

This objection was met by Zermelo who gave a formal proof of the well-ordering principle. His proof, as one might guess from looking at Cantor's argument, involves the assumption of the axiom of choice (see §6.8). It turns out, in fact, that the well-ordering principle is equivalent to the axiom of choice – i.e. either can be deduced from the other.

Let us now recall for what reason all this discussion was initiated. It began with our desire to 'count' the uncountable. In a sense, the well-ordering principle solves this problem for us.

Suppose that we wish to 'count' the uncountable set S. We first use the well-ordering principle and deduce the existence of a well-ordering of S. Relative to this well-ordering, S will have a first element which we count with 1. We then count the second element with 2 and so on until all the elements of $\mathbb{N}$ have been used up.

But since S is uncountable there will still be a subset T of S which has not yet been counted. But, relative to the given well-ordering, T has a first element. This is clearly the next thing to be 'counted'. But with what? All our counting symbols (i.e. the elements of $\mathbb{N}$) have been used up.

Obviously we need more 'counting symbols'. These 'counting symbols' are called the *ordinal numbers*.

12.24†   Ordinal numbers

The finite ordinals are 0, 1, 2, 3 and so on. These may be defined by

$$1 = \{0\}$$
$$2 = \{0, 1\}$$
$$3 = \{0, 1, 2\}$$
$$4 = \{0, 1, 2, 3\}$$
$$\cdots$$

(Recall §10.3. It is usual incidentally, to identify 0 with the empty set $\emptyset$.) But there is no reason to stop with the finite ordinals and we go on to define

$$\omega = \{0, 1, 2, 3, \ldots\}$$
$$\omega + 1 = \{0, 1, 2, 3, \ldots; \omega\}$$
$$\omega + 2 = \{0, 1, 2, 3, \ldots; \omega, \omega + 1\}$$
$$\cdots$$
$$\omega 2 = \{0, 1, 2, 3, \ldots; \omega, \omega + 1, \omega + 2, \ldots\}$$
$$\omega 2 + 1 = \{0, 1, 2, 3, \ldots; \omega, \omega + 1, \omega + 2, \ldots; \omega 2\}$$
$$\cdots$$
$$\omega = \{0, 1, 2, 3, \ldots; \omega, \omega + 1, \omega + 2, \ldots; \omega 2, \omega 2 + 1, \ldots\}$$
$$\omega 3 + 1 = \{0, 1, 2, 3, \ldots; \omega, \omega + 1, \omega + 2, \ldots; \omega 2, \omega 2 + 1, \ldots; \omega 3\}$$
$$\cdots$$

Then, when all the possibilities involving ω, $\omega2$, $\omega3$ and so on are used up, we go on and define

$$\omega^2 = \{0, 1, 2, \ldots; \omega, \omega+1, \ldots; \omega2, \omega2+1, \ldots; \omega3, \omega3+1, \ldots; \ldots; \ldots; \ldots\}.$$

We can then continue with $\omega^2 + 1$ and so on. In general, each ordinal is defined to be the set consisting of all previous ordinals.

Now, we see what to do when trying to 'count' an uncountable set S. When the symbols 1, 2, 3,... have been used up, we 'count' the first uncounted element of S with ω and then the next uncounted element of S with $\omega+1$ and so on until all the symbols ω; $\omega+1$, $\omega+2$,... have been exhausted. Then we move on to $\omega2$ and so on.

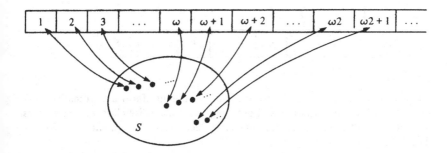

The important point to notice is this. When one well-orders a set S one imposes upon it a certain well-ordering structure. To each structure there corresponds exactly one ordinal number. For example, the two well-orderings of the set $\mathbb{N}$ given by

$$1 \lhd 3 \lhd 5 \lhd 7 \ldots 2 \lhd 4 \lhd 8 \ldots$$

and

$$2 \lhd 4 \lhd 8 \lhd 10 \ldots 1 \lhd 3 \lhd 5 \ldots$$

have the same structure and this structure corresponds to that of the ordinal $\omega2$.

When one writes 1, 2, 3,...; ω, $\omega+1$,...; ... one is therefore essentially listing all the different possible well-ordering structures in increasing order of complexity. Now suppose that S is a given set which we have well-ordered. We can then ascend the scale of the ordinal numbers until we arrive at the ordinal number α which corresponds to the order structure imposed on S by our well-ordering. We shall then have 'counted' the set S and the result of our 'counting' will be the ordinal number α.

For the reader who may feel, with some justice, that the above explanation is ordered in terms of increasing vagueness, we suggest the book *The Theory of Sets and Transfinite Arithmetic* by A. Abian (Saunders, 1965). In this book will be found a formal definition of ordinal numbers and a formal derivation of their properties.

A less precise, but more readable, text is the book *Theory of Sets* by E. Kamke (Dover, 1950).

12.25† Cardinals

Recall that ordinals were defined as certain sets. It therefore makes sense to say that two ordinals have the same cardinality. For example,

$$\omega = \{0,\ 1,\ 2,\ 3,\ldots\}$$

and
$$\omega 2 = \{0,\ 1,\ 2,\ 3,\ldots;\ \omega,\ \omega+1,\ \omega+2,\ldots\}$$

have the same cardinality because of the bijection indicated below.

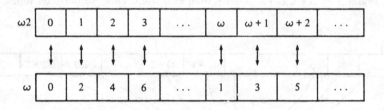

Let us say that all ordinals with the same cardinality form a 'cardinality class'. Since the ordinal numbers are well-ordered, each cardinality class will have a first element. We call the first ordinal number in each class the *cardinal number* of the class.

When we have finished counting a set and arrived at an ordinal number α, we can then sensibly say that the number of elements in the set (i.e. the *cardinal number* of the set) is the smallest ordinal with the same cardinality as α.

We know that there is only one ordinal with four elements – namely the ordinal

$$4 = \{0,\ 1,\ 2,\ 3\}.$$

Hence 4 goes into a cardinality class on its own. It is therefore the first element in this class and so 4 is a cardinal number. If asked to count the set $\{a,\ b,\ c,\ d\}$ we would arrive at the ordinal 4. The smallest ordinal with the same cardinality as 4 is again 4. Hence $\{a,\ b,\ c,\ d\}$ has 4 elements. Similar reasoning applies to all the finite ordinals $1, 2, 3, \ldots$.

So far this is all much ado about nothing since we coped quite well in §12.1 with finite sets without ever mentioning ordinals and cardinals as such. The interest lies in what happens when we look at infinite sets.

The first cardinality class after the finite ones is the class of countably infinite ordinals – i.e. the class of those ordinals which contain an infinite but countable number of elements. Into this class go the ordinals $\omega,\ \omega+1,\ldots;\ \omega 2,\ \omega 2+1,\ldots;\ \ldots$ and many, many more.

The first element of this class is ω and so ω is the first *transfinite* cardinal. When thinking of ω as such we denote it by $\aleph_0$. ($\aleph$ is the Hebrew letter A, pronounced 'aleph'.) Thus the number of elements in the set $\mathbb{N}$ is $\aleph_0$. Or, what is the same thing, all countable sets have cardinal number $\aleph_0$.

In example 12.9, we saw that each element of $\mathbb{N} \times \mathbb{N}$ can be thought of as specifying a square in an infinite array of squares. Suppose that we 'count' these squares as indicated below

1	2	3	4	5
ω	$\omega+1$	$\omega+2$	$\omega+3$	
$\omega 2$	$\omega 2+1$	$\omega 2+3$	$\omega 2+4$	
$\omega 3$	$\omega 3+1$			

The ordinal number at which we arrive is ω^2 and the smallest ordinal with the same cardinality is ω.

We can now consider the cardinal number $\aleph_1$. This is the first transfinite ordinal which does not have cardinality $\aleph_0$ – i.e. it is the first uncountable cardinal.

We can then go on to consider $\aleph_2, \aleph_3, \ldots$ and even $\aleph_\omega, \aleph_{\omega+1}, \ldots$ and so on.

12.26† Continuum hypothesis

We have seen that the set $\mathbb{R}$ of real numbers is uncountable. It is natural to ask whether the cardinal number of $\mathbb{R}$ is $\aleph_1$. Or, to put it another way, does $\mathbb{R}$ have uncountable subsets whose cardinality is not the same as $\mathbb{R}$?

It is not hard to see (by looking at the binary expansions of the real numbers) that the cardinality of $\mathbb{R}$ is the same as that of $2^\mathbb{N}$ and hence that of $\mathscr{P}(\mathbb{N})$. (See exercise 12.22(2).) This leads us to use the notation

$$2^{\aleph_0}$$

for the cardinal number of $\mathbb{R}$ (also called the cardinal of the continuum). The *continuum hypothesis* is that

$$\aleph_1 = 2^{\aleph_0}.$$

The generalised continuum hypothesis is that

$$\aleph_{\alpha+1} = 2^{\aleph_\alpha}$$

for each ordinal α.

Many mathematicians tried to prove this result but without success. It was finally shown (Gödel doing one part of the work and Cohen the other), that the generalised continuum hypothesis is *independent* of the axioms of formal set theory – i.e. it is possible to construct models of set theory in which the hypothesis is true and models in which it is false. Its status is therefore similar to that of the parallel postulate of Euclidean geometry (see §10.1).

The question then arises as to whether or not the continuum hypothesis should be taken as another axiom of set theory. No consensus has yet emerged amongst mathematicians on this issue. In his book *Set Theory and the Continuum Hypothesis* (Benjamin, 1966), P. Cohen gives his view that this would be intuitively unsatisfying and we can do little more than to take his word for it.

NOTATION

INDEX